Lecture Notes in Statistics

Edited by D. Brillinger, S. Fienberg, J. Gani, J. Hartigan, J. Kiefer, and K. Krickeberg

5

Tomasz Rolski

Stationary Random Processes Associated with Point Processes

Springer-Verlag
New York Heidelberg Berlin

Tomasz Rolski
Mathematical Institute
Wroclaw University
pl. Grundwaldzki 2/4
50-384 Wroclaw
Poland

AMS Subject Classification: 60G55, 60K35, 62M99

Library of Congress Cataloging in Publication Data

Rolski, Tomasz.
Stationary random processes associated with point processes.

(Lecture notes in statistics; v. 5)
Bibliography: p.
Includes index.
1. Stationary processes. 2. Point processes.
I. Title. II. Series: Lecture notes in statistics (Springer-Verlag); v. 5.
QA274.3.R64 519.2'32 81-1056
AACR2

Printed in the United States of America

9 8 7 6 5 4 3 2 1

ISBN 0-387-90575-8 Springer-Verlag New York Heidelberg Berlin
ISBN 3-540-90575-8 Springer-Verlag Berlin Heidelberg New York

PREFACE

In this set of notes we study a notion of a random process associated with a point process. The presented theory was inspired by queueing problems. However it seems to be of interest in other branches of applied probability, as for example reliability or dam theory. Using developed tools, we work out known, aswell as new results from queueing or dam theory. Particularly queues which cannot be treated by standard techniques serve as illustrations of the theory.

In Chapter 1 the preliminaries are given. We acquaint the reader with the main ideas of these notes, introduce some useful notations, concepts and abbreviations. We also recall basic facts from ergodic theory, an important mathematical tool employed in these notes. Finally some basic notions from queues are reviewed.

Chapter 2 deals with discrete time theory. It serves two purposes. The first one is to let the reader get acquainted with the main lines of the theory needed in continuous time without being bothered by technical details. However the discrete time theory also seems to be of interest itself. There are examples which have no counterparts in continuous time.

Chapter 3 deals with continuous time theory. It also contains many basic results from queueing or dam theory.

Three applications of the continuous time theory are given in Chapter 4. We show how to use the theory in order to get some useful bounds for the stationary distribution of a random process.

In Chapter 5 we apply the results worked out to nonstandard single server queues.

At the end of each chapter general references are given. Within a chapter we give references only when a technical detail is omitted in the text. As these notes do not aim to give a full account of the theory it is likely that some references have been left out. The theorems, propositions, lemmas and definitions are numbered independently, but a corollary is given the same number as the theorem to which it is a corollary.

I am grateful to Prof.Kopociński fo his valuable comments, to Dr. Schmidt for his critical remarks on Section 3.5, and to Miss Bochynek for typing the manuscript.

Contents

Chapter 1. Preliminaries

§ 1. Introduction.

In applied probability one frequently meets random processes which are associated with point processes. This is a typical situation in queueing theory. For example the queue size process or the virtual waiting time process are associated with the point process of instants of arrival.

Let $X^o = \{X^o(t),\ t\geq 0\}$ denote a random process under investigation and N^o a point process associated with X^o. Assume that both are defined on a probability space (A, SA, Pr) and X^o takes values in some function space $L(R_+, E)$. Let $\{Y_i^o,\ i=0,1,\ldots\}$ denote consecutive positions of points of N^o on the half-line R_+. In these notes we set up and try to give an answer to the following problems:

(i) the stability of X^o,

(ii) a construction of a stationary version of (X^o, N^o),

(iii) relations between a stationary version of $\{X^o(t),\ t\geq 0\}$ and a stationary version of $\{X^o(Y_i^o),\ i=0,1,\ldots\}$.

A few words of comment and explanation as to what we actually mean by those three problems.

Among different approaches to the concept of stability we choose the most natural regarding mathematical methods employed in these notes, i.e. the ergodic theory. Namely it is said that a random process $\{X^o(t),\ t\geq 0\}$, is stable if

$$\lim_{t\to\infty} \frac{1}{t} \int_0^t Pr(\{X^o(u+s),\ u\geq 0\} \in \cdot)ds \tag{1.1}$$

is a proper distribution on $L(R_+, E)$. Sometimes a stronger form of stability can be proved namely that

$$\lim_{s\to\infty} Pr(\{X^o(u+s),\ u\geq 0\} \in \cdot) \tag{1.2}$$

is a proper distribution on $L(R_+, E)$. Then it is said that X^o is strongly stable. The latter case is what we usually understand by saying that X^o is stable and we must be aware of this difference. The definition of stability used here seems to be also justified by our intuition and is useful in application. Similar definitions can be stated in the discrete case. Furthermore notice that if X^o is stable then

$$\lim_{t\to\infty} \frac{1}{t} \int_0^t \Pr(X^o(s) \in \cdot)ds$$

is a proper distribution on E while if X^o is strongly stable then

$$\lim_{t\to\infty} \Pr(X^o(t) \in \cdot)$$

is a proper distribution on E.

Now we begin a discussion of (ii). Denote the limiting distribution in (1.1) by P. A random process having the distribution P is clearly stationary. By stationary we mean that the distribution P is invariant regarding the shift transformation on $L(R_+,E)$. This will be clarified later on. Having a distribution P we can easily construct a random process X with the distribution P. For example we can consider the coordinate random process $\{X(d,t) = d(t),\ t\geq 0\}$ on $(L(R_+,E), SL(R_+,E), P)$. However the main problem has not yet been mentioned. Namely how to determine P having the distribution of (X^o,N^o). It turns out that in some cases it is possible to give an explicit formula for P. This is when $Y_0^o = 0$, a.e. and the distribution of (X^o,N^o) is identical with the distribution of $(\{X^o(Y_1^o + t),\ t\geq 0\}, \{Y_i^o - Y_1^o,\ i=1,2,\dots\})$. Then the distribution P^o of (X^o,N^o) is called a Palm distribution and (X^o,N^o) is called a Palm version. There is a one-to-one correpondence between a Palm distribution P^o and the respective stationary distribution P.

Another important question is when the stability of $\{X^o(Y_i^o), i=0,1,\dots\}$ implies the stability of $\{X^o(t),\ t\geq 0\}$. This kind of question is related to problem (iii). The next very important issue related to problem (iii) derived in these notes is as follows. Let $L(R_+,E)$ now be a space of right-continuous functions with limits from the left. We suppose that (X^o,N^o) is a Palm version and (X,N) is the stationary process corresponding to (X^o,N^o). Denote by $v(\cdot)$ the distribution of $X(t)$ and set

$$v^+(\cdot) = \Pr(X^o(Y_i^o) \in \cdot) ,$$

$$v^-(\cdot) = \Pr(X^o(Y_i^o - o) \in \cdot) .$$

The distribution v^+ is the distribution of the embedded chain at points of N and v^- is the stationary distribution of the embedded chain just before points of N. We shall see that v, v^-, v^+ are related and shall find these relationships. They are called here first and second type relations respectively. To solve the problem we essentially use the relation between a Palm distribution and the stationary distribution. Such an approach to the problem makes understanding considerably easier.

Simultaneously we develop the general theory of random processes associated with point processes, and we apply it in queueing theory. Particularly we focus attention on queues without the independence assumption about the inter-arrival times and service times. A theory of such processes is outlined here. We do not aim, however, to give a full account of such queues. We deal only with the problem of stability, Takács relations and the Little formula for such queues. Recall that Takács relations relate the distribution (univariate) of the stationary virtual waiting time with the distribution (univariate) of the stationary actual waiting time. The Little formula relates the mean queue size with the mean actual waiting time (both stationary).

We apply general results to special queues, for example to single server queues with work-conserving disciplines and many-server queues with "first in first out" (FIFO) disciplines. We also work with queues with nonhomogeneous processes of the instants of arrival e.g. with a "periodic" in some sense input process or with queues with "fagging" servers. The common feature of all the queues just mentioned is that they cannot be described by standard techniques used in queues with independent inter-arrival and service times.

As we mentioned some parts of these notes are devoted to Takács relations. It will become clear that they are in fact the first and second type relation respectively. In this case as a random process we consider the virtual waiting time process, and as an associated point process we take the point process of instants of arrival. Then, assuming the "first in first out" discipline, the virtual waiting time process at instants just before arrival is the actual waiting time process.

§ 2. Some notations and conventions

Sets of numbers:

$N = \{1,2,\dots\}$,

$N_0 = \{0,1,\dots\}$,

$Z = \{\dots,-1,0,1,\dots\}$,

$R = (-\infty,\infty)$,

$R_+ = [0,\infty)$.

Sequences: a sequence of elements we denote either by $\{x_i, i\in T\}$, or by $\{x_i\}_{i\in T}$. If $T = Z$ (or $T = R$) then we write usually simply $\{x_i\}$ instead of $\{x_i, i\in Z\}$ (or $\{x_t\}$ instead of $\{x_t, t\in R\}$) .

Product of sets: we denote $E^l = \mathop{X}\limits_{i=1}^{l} E$. For example R^l is the l-dimensional Euclidean space. The family of all E-valued functions on T we denote by E^T. Thus $E^T = \mathop{X}\limits_{i=-\infty}^{\infty} E$.

σ-fields: for a given space A we denote the respective σ-field by $S(A)$ or SA. If A is a topological space then the σ-field of Borel subsets is denoted by BA. The Kolmogorov σ-field of subsets of E^T is the σ-field generated by sets of the form

$$\{\{e_\tau,\tau\in T\},\ e_{\tau_i}\in E_i,\ i=1,\dots,k\},\qquad k\in N,\ \tau_i\in T,\ E_i\in SE.$$

If E is metric, $T = Z$ then the Kolomogorov σ-field is a Borel σ-field.

Distribution functions and moments: if F is a distribution function on (R,BR) then

$$m_F = \int_R x\,F(dx) ,$$

$$m_{F,k} = \int_R x^k\,F(dx) ,$$

$$\tilde F(x) = \int_0^x (1 - F(t))dt/m_F ,$$

$$M_m(x) = 1 - \exp(-x/m) \quad \text{if} \quad x\ge 0 \quad \text{and} \quad 0 \quad \text{if} \quad x<0 .$$

Other functions:

$$1_E(x) = 1 \quad \text{if} \quad x\in E \quad \text{and} \quad 0 \ \text{if} \quad x\notin E ,$$

$$\delta_a(x) = 1 \quad \text{if} \quad x \ge a \quad \text{and} \quad 0 \quad \text{if} \quad x < a \ ,$$

$$\delta_a(A) = 1_A(a) \ , \qquad a \in A \ , \qquad A \in \mathcal{S}A$$

$$\# A = \text{the cardinal number of} \ A \ .$$

Random elements: A mapping $\Phi\colon (A,\mathcal{S}A,\Pr) \to (L,\mathcal{S}L)$ is called a *random element*. The probability measure $\Pr \Phi^{-1}$ on $(L,\mathcal{S}L)$ is called the *distribution of* Φ. If $L = R$ then Φ is called a random variable. Then we denote

$$E\,\Phi = \int_A \Phi(a)\Pr(da) \ .$$

Space $\mathcal{D}(I,E)$: Let E be a Polish space and I be a subinterval of the real line. The endpoints of I can be finite or infinite and if finite, open or closed. The space $\mathcal{D}(I,E)$ is the family of all E-valued right-continuous functions on I with limits from the left. The space $\mathcal{D}(I,E)$ with the Skorohod's topology is Polish (i.e. there exists some separable and complete metric generating the topology). Moreover the Kolmogorov σ-field is a Borel σ-field. The projection π_t defined by $\pi_t d = d(t)$, $d \in \mathcal{D}(I,E)$ is measurable.

Space $C(I,E)$: Let I and E be as formerly. The space $C(I,E)$ is the family of all E-valued continuous functions on I. It is a measurable subset of $\mathcal{D}(I,E)$.

Space $N(E)$: Let E be a locally compact, second countable Hausdorff topological space. The space $N(E)$ is the family of all nonnegative integer valued Radon measures on E. The space $N(E)$ with the vague topology is Polish and the Borel σ-field is generated by sets of the form

$$\{n,\ n(E) = k\} \ , \qquad k \in N_0, \quad E \in \mathcal{B}E \ .$$

Abbreviations:

r.v. - random variable,
r.e. - random element,
r.p. - random process,
p.p. - point process,
m.p.p. - marked point process,
d.f. - distribution function,
i.i.d. - independent, identically distributed,

r.p. @ p.p. - random process associated with point process
v.w.t.p. - virtual waiting time process,
a.w.t.p. - actual waiting time process,
r.p. @ m.p.p. - random process associated with marked point process.

§ 3. Ergodic theory - discrete parameter.

We recall some fundamental notions from ergodic theory. Let (A, SA, Pr) be a probability space and $\theta: A \to A$ be measurable.

Definition 1.1:

θ is called *measure-preserving* if $Pr(\theta^{-1}A) = Pr(A)$, $A \in SA$.

Definition 1.2:

A set $A \in SA$ is *invariant* if $\theta^{-1}A = A$.

It is not difficult to prove that the class of invariant sets is a σ-field. We denote it by I_θ.

Definition 1.3:

θ is called *ergodic* if for every $A \in I_\theta$, $Pr(A) = 0$ or 1. If τ is measure-preserving and ergodic then θ is called *metrically-transitive*.

Proposition 1.1:

Let θ be measurable on (A, SA) and define M as the set of all probabilities Pr on A, such that θ is measure-preserving on (A, SA, Pr). Then M is convex and the extreme points of M are the probabilities Pr such that θ is ergodic on (A, SA, Pr).

One of the most important results is the result usually referred to as the *ergodic theorem*.

Theorem 1.1:

Let θ be measure-preserving on (A, SA, Pr). Then for X any r.v. such that $E|X| < \infty$,

$$\lim_{j\to\infty} \frac{1}{j} \sum_{i=0}^{j-1} X(\theta^i a) = E(X|I_\theta) \text{ , a.e.}$$

Moreover if θ is ergodic then

$$\lim_{j\to\infty} \frac{1}{j} \sum_{i=0}^{j-1} X(\theta^i a) = EX \text{ , a.e..}$$

§ 4. Ergodic theory - continuous parameter

Let (A,SA,Pr) be a probability space and $\underline{\theta} = \{\theta^t,\ t \in R\}$ be a group of one-to-one measurable transformations of A into itself, i.e.

(i) θ^0 is the identity transformation,

(ii) $\theta^{t+s} = \theta^t\theta^s$, $t,s \in R$.

Definition 1.4:

θ is called *measure-preserving* if for any $t \in R$ the transformation is measure-preserving. If the mapping

$$R \times A \ni (t,a) \to \theta^t a \in A$$

is measurable then $\underline{\theta}$ is called *measurable*.

Definition 1.5:

A set $A \in SA$ is *invariant* if $\theta^t A = A$, any $t \in R$.

The class of invariant subsets is a σ-field. We denote it by $I_{\underline{\theta}}$.

Definition 1.6:

$\underline{\theta}$ is called *ergodic* if for every $A \in I_{\underline{\theta}}$, $Pr(A) = 0$ or 1. If $\underline{\theta}$ is measure-preserving and ergodic then $\underline{\theta}$ is called *metrically-transitive*.

The next result is usually referred to as the *continuous parameter ergodic theorem*.

Theorem 1.2:

Let $\underline{\theta}$ be a measurable group of measure-preserving transformations of A into itself. Then for any r.v. X such that $E|X| < \infty$

$$\lim_{t\to\infty} \frac{1}{t} \int_0^t X(\theta^s a)ds = E(X|I_{\underline{\theta}}) \text{ , a.e..}$$

Moreover if $\underline{\theta}$ is ergodic then

$$\lim_{t\to\infty} \frac{1}{t} \int_0^t X(\theta^s a)ds = EX \ , \quad \text{a.e.}$$

§ 5. Stationary random elements

Let $\Phi: (A,SA,Pr) \to (G,SG)$ be an r.e. There is given on G a measurable group $\underline{\theta} = \{\theta^t, t\in T\}$ of transformations of G into itself. In this set of notes $T = Z$ or R.

Definition 1.7:

The r.e. Φ is called *stationary* if the group $\underline{\theta}$ is measure-preserving on $(G,SG,Pr\,\Phi^{-1})$. It is said that Φ is *metrically-transitive* if $\underline{\theta}$ is metrically-transitive on $(G,SG,Pr\,\Phi^{-1})$. A probability measure P on G is called *stationary* if $P\theta^t = P$, $t\in T$.

Definition 1.8:

An r.e. $\Phi: (A,SA,Pr) \to (G,SG)$ is called *stable* if for any $G \in G$

(1.3) $$\lim_{j\to\infty} \frac{1}{j} \sum_{i=0}^{j-1} Pr(\theta^i\Phi \in G) = P(G) \quad (\text{if } T = Z)$$

or

(1.3') $$\lim_{t\to\infty} \frac{1}{t} \int_0^t Pr(\theta^s\Phi \in G)ds = P(G) \quad (\text{if } T = R)$$

exists and is an honest distribution on G. The r.e. Φ is called *strongly stable* if for any $G\in SG$

$$\lim_{t\to\infty} Pr(\theta^t\Phi \in G) = P(G) \quad (t\in Z \text{ or } R \text{ respectively})$$

exists and P is an honest distribution on G.

Clearly $\underline{\theta}$ is a measure-preserving transformation on (G,SG,P) when P is given by (1.3) or (1.3') respectively. The distribution P is called the *stationary distribution* of the r.e. Φ. If $\{X_t, t\in T\}$ is a sequence of random vectors (i.e. $X_t \in R^1$, $t \in T$) then

$$F(x) = \lim_{j\to\infty} \frac{1}{j} \sum_{i=0}^{j-1} Pr(X_i \le x) ,$$

or respectively

$$F(x) = \lim_{t\to\infty} \frac{1}{t} \int_0^t Pr(X_s \le x)ds , \qquad x \in R^1$$

is called the *stationary distribution function* of $\{X_t, t \in I\}$. If two stable r.e.'s $\Phi_i: (A,SA,Pr) \to (G,SG)$, $i=1,2$ have the same distribution then we call them *equivalent*.

The following proposition is frequently used in this set of notes. In terms of r.e.'s its meaning is such that having $\underline{\theta}_1$-stationary ($\underline{\theta}_1$-metrically-transitive) r.e. Φ_1 we find when the r.e. $\Phi_2 = \varphi(\Phi_1)$ is $\underline{\theta}_2$-stationary ($\underline{\theta}_2$-metrically-transitive).

Proposition 1.2:

Let (G_i, SG_i), $i=1,2$ be measurable spaces and $\underline{\theta}_i$, $i=1,2$ be groups of transformations of G_i, $i=1,2$ into itself respectively. Let

$$\varphi: G_1 \to G_2$$

be a measurable mapping.
Assume that for any $\theta_2 \in \underline{\theta}_2$ there exists $\theta_1 = \psi(\theta_2) \in \underline{\theta}_1$ such that

$$\varphi\theta_1 = \theta_2\varphi \tag{1.4}$$

and the mapping ψ is surjective, i.e.

$$\psi(\underline{\theta}_2) = \underline{\theta}_1 . \tag{1.5}$$

If an r.e.

$$X: (A,SA,Pr) \to (G_1,SG_1)$$

is stationary (metrically-transitive) then the r.e. $Y = \varphi(X)$ is stationary (metrically-transitive).

Proof: For any $E_2 \in SG_2$ and $\theta_2 \in \underline{\theta}_2$

$$Pr(\theta_2 Y \in E_2) = Pr(\theta_2\varphi(X) \in E_2) =$$

$$= Pr(\varphi(\theta_1 X) \in E_2) = Pr(\theta_1 X \in \varphi^{-1}E_2) =$$

$$= Pr(X \in \varphi^{-1}E_2) = Pr(Y \in E_2) .$$

Thus Y is stationary. Now assume that X is ergodic. Thus for any $\underline{\theta}_1$-invariant set A

$$\Pr X^{-1}(A) = 0 \text{ or } 1 .$$

Now let B be a $\underline{\theta}_2$-invariant set. We can demonstrate that $\varphi^{-1}B$ is $\underline{\theta}_1$-invariant. We have $\theta_2^{-1}B = B$, any $\theta_2 \in \underline{\theta}_2$, which yields

$$(\theta_2\varphi)^{-1}B = \varphi^{-1}B .$$

Thus there exists $\theta_1 \in \underline{\theta}_1$ such that

(1.6) $$\theta_1^{-1}\varphi^{-1}B = \varphi^{-1}B .$$

Because of (1.5) we have that (1.6) holds for any $\theta_1 \in \underline{\theta}_1$. Hence it follows that $\varphi^{-1}B$ is a $\underline{\theta}_1$-invariant set.
To complete the proof we note that

$$\Pr Y^{-1}(B) = \Pr(\varphi(X) \in B) = \Pr X^{-1}(\varphi^{-1}B) = 0 \text{ or } 1 .$$

Let E be a Polish space and $X = \{X(t),\ t\in T\}$ be an r.p. with the state space E defined on a probability space $(A,SA,\Pr)$. In probability theory the r.p. X is called *stationary* if for any $t,t_1,\dots,t_k \in T$, $E_1,\dots,E_k \in BE$

$$\Pr(X_{t_1} \in E_1,\dots,X_{t_k} \in E_k) = \Pr(X_{t+t_1} \in E_1,\dots,X_{t+t_k} \in E_k) .$$

The r.p. X can be considered as an r.e. If $T = \mathrm{R}$ then we assume that trajectories of X belong to $D(\mathrm{R},E)$. Define the group of transformation $\underline{\tau}_*$ by

$$\tau_*^t d = d(\cdot + t), \qquad t \in T ,\quad d \in L ,$$

where now $L = E^{\mathrm{Z}}$ or $D(\mathrm{R},E)$ respectively. The group of transformations $\underline{\tau}_*$ is measurable and measure-preserving on $(L,BL,\Pr X^{-1})$. The measure-preserving property is clear while the measurability follows easily from the following lemma.

<u>Lemma</u> 1.1.
The mapping

$$\mathrm{R} \times D(\mathrm{R},E) \ni (t,d) \to \tau_*^t d \in D(\mathrm{R},E)$$

is continuous.

Proof: For a fixed $t \in R$ denote $g_t(s) = t+s$, $s \in R$. Then $\tau_*^t d(s) = d(s+t) = dg_t(s)$, $s \in R$. Whitt (1979), Theorem 3.1 proved that the mapping

$$\mathcal{D}(R,E) \times C(R,R) \ni (d,g_t) \to dg_t \in \mathcal{D}(R,E)$$

is continuous. The family $\{g_t(\cdot),\ t \in R\}$ is closed in $C(R,R)$ and is homeomorphic with R. Hence the mapping $\tau_*^t d$ is continuous.

In these notes we shall frequently use the result of the next lemma for proving the stability of an r.p. Denote $T_+ = R_+$ or N_0 and $L = \mathcal{D}(R,E)$ or E^Z respectively.

Lemma 1.2:

Assume that there exists a stable (strongly stable) r.p. $\{X(t), t \in T\}$ with the distribution P and a T_+-valued r.v. Z such that

$$Y(Z+t) = X(Z+t), \quad t \in T_+ .$$

Then Y is stable (strongly stable) with the stationary distribution P.

Proof: We demonstrate the proof in the case $T = R$. The case $T = Z$ is identical. We have that $B\mathcal{D}(R,E)$ is generated by the family

$$\bigcup_{i=-1}^{-\infty} B\mathcal{D}([i,\infty),E) ,$$

where $B\mathcal{D}([i,\infty),E)$ should be now understood as the σ-field generated by sets

$$\{d \in \mathcal{D}(R,E):\ d(t_j) \in B_j,\ j=1,\dots,k\}, \quad t_j \geq i,\ B_j \in BE,\ k=1,2,\dots .$$

Let $D \in \mathcal{D}([i,\infty),E)$. Then

$$\lim_{t\to\infty} \frac{1}{t} \int_0^t \Pr(\tau_*^s Y \in D)\,ds =$$

$$= \lim_{t\to\infty} \frac{1}{t} \int_0^t (\Pr(\tau_*^s Y \in D,\ s \geq Z+i) + \Pr(\tau_*^s Y \in D,\ s < Z+i))\,ds =$$

$$= \lim_{t\to\infty} \frac{1}{t} \int_0^t (\Pr(\tau_*^s X \in D,\ s \geq Z+i) + \Pr(\tau_*^s Y \in D,\ s < Z+i))\,ds .$$

Since Z is an r.v. we have

$$\lim_{t\to\infty} \Pr(Z+i<t) = 1$$

which yields

$$\lim_{t\to\infty} \frac{1}{t}\int_0^t \Pr(\tau_*^s X \in D,\ Z+i<s)ds = P(D) .$$

We also have

$$\lim_{t\to\infty} \Pr(Z+i\geq t) = 0$$

which yields

$$\lim_{t\to\infty} \frac{1}{t}\int_0^t \Pr(\tau_*^s Y \in D,\ Z+i\geq s)ds = 0 .$$

Hence

$$\lim_{t\to\infty} \frac{1}{t}\int_0^t \Pr(\tau_*^s Y \in D)ds = P(D) . \tag{1.7}$$

Now we show that (1.7) holds for any $D \in \mathcal{BD}(R,E)$. Define

$$\mathcal{D}_0 = \{D \in \mathcal{D}(R,E):\ \lim_{t\to\infty} \frac{1}{t}\int_0^t \Pr(\tau_*^s Y \in D)ds = P(D)\} .$$

It is clear that any $D \in \bigcup_{i=-1}^{\infty} \mathcal{BD}([i,\infty),E)$ belongs to $\mathcal{D}_0$. Thus to complete the proof it would suffice to find out that $\mathcal{D}_0$ is a monotone class. Let $D_1 \subset D_2 \subset \ldots \in \mathcal{D}_0$. Set $D = \bigcup_{i=1}^{\infty} D_i$. There exists j_0 such that

$$P(\bigcup_{i=1}^{j} D_i) \geq P(D) - \frac{\varepsilon}{2} , \qquad j \geq j_0 .$$

There also exists $t_0 > 0$ such that

$$\frac{1}{t}\int_0^t \Pr(\tau_*^s Y \in \bigcup_{i=1}^{j_0} D_i)ds \geq P(\bigcup_{i=1}^{j_0} D_i) - \frac{\varepsilon}{2} , \qquad t \geq t_0 .$$

Hence

$$\lim_{t\to\infty} \frac{1}{t}\int_0^t \Pr(\tau_*^s Y \in D)ds \geq P(D) .$$

Since $\mathcal{D}_0$ is closed under the complement operation, we have that $\mathcal{D}_0$ is a monotone class and hence $\mathcal{D}_0 \supset \mathcal{D}(R,E)$. This completes the proof of the lemma.

§ 6. Loynes' lemma

In applied probability, frequently system characteristics of interest can be generated recursively. This is a question when such characteristics are stable. In some cases an answer is given by Loynes' lemma (see Loynes (1962)).

Lemma 1.3:

Let random elements $\underline{X}_i$: $(A,AS,Pr) \to (R_+^k, \mathcal{B}(R_+^k)), i \in N_0$ be related by transformation

(1.8) $$\underline{X}_{i+1} = f(\underline{X}_i, U_i),$$

where U_i: $(A,SA,Pr) \to (E,SE)$ form a metrically-transitive sequence. Suppose that f: $R_+^k \times E \to R_+^k$ is increasing and continuous from the left (even at $(\infty,\dots,\infty)$) with respect to the first variable. Then there exists a sequence (possibly dishonest) of random vectors $\{\underline{M}_i\}$ satisfying

$$\underline{M}_{i+1} = f(\underline{M}_i, U_i)$$

such that the sequence $\{\underline{M}_i, U_i\}$ is metrically-transitive. If $\underline{X}_0 = \underline{0} = (0,\dots,0) \in R^k$, then the distribution functions of $(\underline{X}_i, \underline{X}_{i+1}, \dots, \underline{X}_{i+k})$ tend monotonically to that of $(\underline{M}_0, \dots, \underline{M}_k)$ as i tends to ∞. Furthermore $\{\underline{M}_i\}$ is the minimal sequence satisfying (1.8) for all i.

Actually, the statement of the above lemma differs from the original one in stating that $\{\underline{M}_i, U_i\}$, instead of $\{\underline{M}_i\}$, is stationary. We have also strengthened the assertion of the lemma that $\{\underline{M}_i, U_i\}$ is metrically-transitive assuming that $\{U_i\}$ is. The main limes of the proof remain the same as in the original one of Loynes.

§ 7. Queues

The principal system of characteristics of interest are: *actual waiting time* - time from arrival to commencement of service, *virtual waiting time at time* t (*work-load at time* t) - the work still to be done by the system at time t if after t no new customer were to arrive, *sojourn time* - time from arrival to completion of service, *queue size at time* t - the number of customers in the system at time t, *busy period* - interval during which the server is busy (in the case of single server queues), *idle period* - interval during which the server is idle (in the case of single server queues), *busy cycle* - idle period plus the preceding busy period.

In these notes we define a queue by giving a *generic sequence*. The generic sequence carries out all information needed to perform the evolution of the system provided that initial conditions are known. Usually a generic sequence contains inter-arrival times between the consecutive arrivals of customers to the system and service times of consecutive customers. Sometimes customers are devided into a few classes and inter--arrival times as well as service times differ within any class respectively.

It is said that a queue is *work-conserving* if the v.w.t.p. in that system is identical to the v.w.t.p. in the same system but with the FIFO discipline.

For an a.w.t.p. $W = \{W_i\}$ define

$$i_0(W) = \inf\{i: W_i > 0\} .$$

A discipline is called *normal* if there exists a Borel function

$$\phi: (R_+ \times R_+)^N \to R_+^N$$

such that for any a.w.t.p. $W = \{W_i\}$ and for any $\alpha \in \{i > i_0(W): W_i = 0\}$ we have

$$(W_\alpha, W_{\alpha+1}, \ldots) = \phi((T_\alpha, S_\alpha), (T_{\alpha+1}, S_{\alpha+1}), \ldots) , \quad \text{a.e.} .$$

This means that the knowlegde of inter-arrival times and service times in a busy cycle is sufficient to determine waiting times of customers in this busy cycle, except for the first busy cycle in which some initial conditions must be given. In these notes we employ two kinds of queueing disciplines namely

(a) first in first out (FIFO),

(b) work-conserving, normal.

Note that within the class of queues with work-conserving normal disciplines there are among others the disciplines: first in first out, last in last out, service in random order, relative priority disciplines.

A queue with a generic sequence is called *stable* if the a.w.t.p. in the queue is stable.

Now we review two basic types of queues.

(i) G/G/1 *queue*. This queue is of infinite capacity. The generic sequence is

$$\{(T_i, S_i, K_i),\ i \in Z\}$$

and is assumed metrically-transitive. Here T_i denotes the inter-arrival time between the i-th and (i+1)-st customer, S_i denotes the service time of the i-th customer and K_i denotes the class which the i-th customer comes from, $i \in Z$. We assume that

$$T_i > 0\ , \qquad ET_i < \infty\ ,$$

$$K_i \in K \subset Z\ , \qquad i \in Z\ .$$

Assuming the FIFO discipline the actual waiting time W_i of the i-th customer can be generated recursively by

$$W_{i+1} = \max(0, W_i + S_i - T_i)\ . \tag{1.9}$$

From Loynes (1962) we know that there exists a sequence M_i, $i \in Z$ such that (1.9) holds and $\{(M_i, T_i, S_i, K_i),\ i \in Z\}$ is metrically-transitive. If the traffic intensity

$$\rho = \frac{ES_0}{ET_0}$$

is less than 1 then M_i, $i \in Z$ are finite and

$$M_{i+1} = \max(0, S_i - T_i,\ S_i - T_i + S_{i-1} - T_{i-1} + \ldots)\ , \qquad i \in Z\ .$$

However in some cases M_i, $i \in Z$ are finite though $\rho = 1$.

(ii) G/G/k; FIFO *queue*. The queue is of infinite capacity. The generic sequence is

$$\{(T_i, S_i, K_i),\ i \in Z\}$$

and is assumed metrically-transitive. The meanings and assumptions on $\{T_i, S_i, K_i\}$ are the same as in the single server case. In the multiserver case we deal only with the FIFO discipline and such systems are denoted G/G/s; FIFO. In this case the analysis of the queue is best approached through a vector-valued waiting time process $\{\underline{W}_i,\ i \in Z\}$.

We assume the customers are served by the first available server. If more than one server is free, there is a fixed rule of choosing a server. Let $\underline{W}_i = (W_{i1}, \ldots, W_{ik})$ be the vector-valued process whose components W_{ij} denote the work-load of the server with the j-th lightest load just prior to the arrival of the i-th customer. Thus W_{i1} is the actual waiting time of the i-th customer. The following recursive relationship exists for the $\underline{W}_i$'s

$$\underline{W}_{i+1} = [R(\underline{W}_i + \underline{S}_i - \underline{T}_i)]^+, \tag{1.10}$$

where $\underline{S}_i = (S_i, \ldots, 0)$, $\underline{T}_i = (T_i, \ldots, T_i)$, $[\underline{X}]^+ = (\max(0, X_1), \ldots, \max(0, X_k))$ for $\underline{X} = (X_1, \ldots, X_k)$ and $R: R^k \to R^k$ rearranges the components of its arguments in ascending order. From Loynes' lemma we know that there exists a sequence $\underline{M}_k$ satisfying (1.10) such that $\{\underline{M}_i, T_i, S_i, K_i\}$ is metrically-transitive. Moreover if $\underline{W}_0 = \underline{0}$ then the distribution functions of $\underline{W}_i$ tends to that of $\underline{M}_0$ as i tends to infinity. If the traffic intensity

$$\rho = \frac{ES_0}{kET_0}$$

is less than 1 then M_i, $i \in Z$ are finite a.e. .

Other systems will be introduced in the text. Finally we note the following convention used in the notes. If r.v.'s in a component of the generic sequence are i.i.d. then in Kendall's symbol denoting a queue we write GI instead of G respectively. For example GI/GI/1 denotes a single server queue with a generic sequence $\{T_i, S_i\}$ and the two sequences $\{T_i\}$ and $\{S_i\}$ each consist of i.i.d. r.v.'s and are themselves independent while in a GI/G/1 queue the sequence $\{T_i\}$ consists of i.i.d. r.v.'s and is independent of $\{S_i\}$. As usual if a component of the generic sequence consists of i.i.d. r.v.'s with a common negative exponential distribution we denote it by M.

Notes

In a few papers the idea of employing the theory of stationary p.p.'s in investigation of a class of r.p.'s was utilized, among others the paper of Miyazawa (1977) and especially the papers of Franken and his co-workers, e.g. Franken (1976), Franken & Streller (1979), Arndt & Franken (1979). Franken and his co-workers made use of the concept of stationary m.p.p.'s on the real line, in which a mark represents a piece of trajectory of a considered r.p.. In my opinion the concept of r.p.@ p.p. which will be investigated in these notes, is more natural and covers a broader class of models.

The concept of the Palm distribution was originally introduced for stationary p.p.'s. We follow the idea of Ryll-Nardzewski (1961) and Mecke (1967) in this field.

The relationships between v, v^+, v^- were presented in a primitive form by Kuczura (1973) (in the case of a denumerable state space) and Jankiewicz & Rolski (1977) (in the case of a general state space). An interesting method of finding relationships in queues for queue size process was proposed by Kopocińska & Kopociński (1971) (1972). The most general case was first considered by König et al (1978). In the latter paper there is also enclosed an extensive bibliography of earlier papers in this field. In these notes we follow an approach of Rolski (1977a) and König et al (1978) for establishing the mentioned relations.

Takács proved relations between the virtual waiting time and the actual waiting time in his papers (1955), (1963).

Further information about the space $\mathcal{D}(I,E)$ and proofs of the facts mentioned in these notes can be found in the paper of Whitt (1979).

For details about the space $N(E)$ we refer to the book of Kallenberg (1976).

The ergodic theory is an important mathematical tool used in these notes. For details the reader is referred to the books of Breiman (1968) and Doob (1953).

Required information on queues the reader can find in the original paper of Loynes (1962) and in the book of Borovkov (1972). Work conservative queues were considered by Wolf (1970).

CHAPTER 2. DISCRETE TIME r.p.@ p.p.

§ 1. Basic concepts.

In this chapter we study a sequence $\Phi = \{X_i, N_i\}$ defined on a probability space $(A, SA, \Pr)$. We assume that each X_i assumes values in a Polish space E, while $N_i = 0$ or 1, $i \in Z$. Such a sequence is called a *discrete time random process associated with a point process* (r.p.@ p.p.). The second component of Φ namely the r.p. $\{N_i\}$ is called a *point process* (p.p.). If $N_i = 1$ we read that there is a point at i.

Denote

$$F = (E \times \{0,1\})^Z .$$

Clearly F is a Polish space. Notice that

$$\Phi\colon (A, SA, \Pr) \to (F, BF) .$$

Denote $\Phi\hat{} = (X\hat{}, N\hat{})$, where for $\{x_i, n_i\} \in F$

$$X\hat{}_j(\{x_i, n_i\}) = x_j , \qquad N\hat{}_j(\{x_i, n_i\}) = n_j , \qquad j \in Z .$$

Similarly measurable functions on F we provide with the mark $\hat{}$.

For a given $f = \{d_i, n_i\} \in F$, we denote by $\{y_i\}$ the consecutive coordinates of all points of $n = \{n_i\}$ on Z indexed according to the convention

$$\dots < y_0 \le 0 < y_1 < y_2 < \dots .$$

The consecutive inter-point distances we denote by

$$t_i = y_{i+1} - y_i , \qquad i \in Z .$$

Notice that the y_i, $i \in Z$ are defined by the second component n but it is convenient to write that the y_i, $i \in Z$ are functions of $f = (d, n)$. It is easy to see that the mappings

$$\hat{Y_i}(f) = y_i ,$$

$$\hat{T_i}(f) = t_i , \qquad i \in Z$$

are measurable. Thus coordinates of points from N

$$Y_i = \hat{Y_i}(\Phi) ,$$

and inter-point distances between points of N

$$T_i = \hat{T_i}(\Phi) , \qquad i \in Z$$

are r.v.'s.

Denote by

$$F^o = \{\{d_i, n_i\} : n_0 = 1\} .$$

Clearly $F^o \in BF$ and contains all $f = (d,n)$ such that n has a point at zero. We have

$$BF^o = \{F \cap F^o : F \in BF\} .$$

In this chapter our objective is to study stationary processes. Thus we need to introduce shift transformations of F into itself.

Definition 2.1:

On (F, BF) define the *shift transformation* τ by

$$\tau(\{d_i, n_i\}) = \{d_{i+1}, n_{i+1}\}$$

and the *shift transformation* $\sigma: F \to F$ by

$$\sigma(\{d_i, n_i\}) = \{d_{i+y_1}, n_{i+y_1}\} .$$

The transformation σ shifts any element $f \in F$ to the left on the distance between zero and the nearest point to the right of zero. We write $\tau^i = \tau^{i-1}\tau$, $i=1,2,\ldots$, where τ^o is the identity transformation of F and $\sigma^i = \sigma^{i-1}\sigma$, $i=1,2,\ldots$, $\sigma^o = \tau^o$. Notice moreover that τ^i and σ^i, $i \in Z$, are measurable transformations of F into itself. Moreover τ^i maps F one-to-one onto F and σ^i maps F^o one-to-one onto F^o, $i \in Z$. In fact $\underline{\tau} = \{\tau^i\}$ and $\underline{\sigma} = \{\sigma^i\}$ are groups of transformations of F or F^o into itself respectively.

Following Section 1.5 we consider Φ as an r.e. on (A, SA, Pr) which assumes values in F, and F is provided with the transformation groups $\underline{\tau}$ or $\underline{\sigma}$. In these notes we use a general terminology introduced in Section 1.5 adapted to the case of the r.p.@ p.p. Φ. If not otherwise stated F is thought to be provided with the group $\underline{\tau}$. Thus we simply say that Φ is for example stationary or metrically-transitive. In the second case it is said for example that an event A is $\underline{\sigma}$-ergodic.

Definition 2.2:

A distribution P^o on (F, BF) is said to be a *Palm distribution* if

$$P^o \sigma^{-1} = P^o .$$

An r.p.@ p.p. Φ^o is said to be a *Palm version* if its distribution $P^o = Pr(\Phi^o)^{-1}$ is a Palm distribution.

In other words P^o is a Palm distribution if it is a $\underline{\sigma}$-stationary distribution and Φ^o is a Palm version if it is a $\underline{\sigma}$-stationary r.e.

Proposition 2.1:

If P^o is a Palm distribution on F then $P^o(F^o) = 1$.

Proof: Assume that $P^o(F^o) < 1$. Then $P^o(F \setminus F^o) > 0$. However we have

$$P^o(\sigma^{-1}(F \setminus F^o)) = P^o(\emptyset) = 0 .$$

Thus

$$P^o(\sigma^{-1}(F \setminus F^o)) \neq P^o(F \setminus F^o)$$

which contradicts the supposition that P^o is a Palm distribution.

Points from N^o form a set of instants of interest. Later we shall investigate the r.p. $\{X^o(t)\}$ at instants from N^o or just before instants from N^o. Denote

$$X_i^- = X^o_{Y_i^o - 1} ,$$

$$X_i^+ = X^o_{Y_i^o} , \qquad i \in Z .$$

The following proposition is easy to prove due to the assumption that Φ^o is a Palm version.

Proposition 2.2:

$\{X_i^-, X_i^+\}$ is a stationary sequence of r.e.'s on (A, SA, Pr).

Example 2.1 (generalized-regenerative processes):

The r.p.@ p.p. $\Phi^o = \{X_i^o, N_i^o\}$ will be defined as follows. Consider a sequence $\Psi_i = \{T_i^o, Z_{i1}^o, \ldots, Z_{iT_i}^o\}$ of r.e.'s on (A, SA, Pr), where $T_i \in N$, $Z_{ij} \in E$. We assume that $\{\Psi_i\}$ is a stationary sequence of r.e.'s. Each Ψ_i, $i \in Z$ describes the r.p.@ p.p. Φ^o in the i-th cycle. The i-th cycle, $i \in Z$ lasts from Y_i^o up to $Y_{i+1}^o - 1$, where

$$Y_j^o = \begin{cases} \sum_{i=0}^{j-1} T_i^o , & j = 1, 2, \ldots , \\ 0 , & j = 0 , \\ -\sum_{i=-1}^{j} T_1^o , & j = -1, -2, \ldots . \end{cases}$$

The Y_i^o, $i \in Z$ are the coordinates on Z of the p.p. $N^o = \{N_i^o\}$. Namely $N_i^o = 1$ iff for some $i \in Z$ we have $Y_j^o = i$. Within the i-th cycle, i.e. for $Y_i^o \le j < Y_{i+1}^o$ the r.p. $\{X_j^o\}$ is defined by

$$X_j^o = Z^o_{i, Y_i^o + j + 1} .$$

The r.p.@ p.p. Φ^o is called a *generalized-regenerative process*. If Ψ_i, $i \in Z$ are i.i.d. r.e.'s then Φ^o is called a *regenerative process*. It is not difficult to prove that Φ^o is a Palm version. This is due to the assumption that $\{\Psi_i\}$ is a stationary sequence.

§ 2. Construction of stationary processes.

Having an r.p.@ p.p. $\Phi^o = \{X_i^o, N_i^o\}$ on a probability space (A, SA, Pr) we often ask for a value and the existence of

$$(2.1) \qquad \lim_{k \to \infty} \frac{1}{k} \sum_{j=0}^{k-1} Pr(\{X_{i+j}^o, N_{i+j}^o, \ i \in Z\} \in F) = P(F) , \qquad F \in BF .$$

The probability measure P on F is called the *stationary distribution* of Φ^o. An r.p.@ p.p. $\Phi = \{X_i, N_i\}$ (possible on another probability space than (A, SA, Pr)) with the distribution P is called a *stationary* r.p.@ p.p. *corresponding to* Φ^o. Clearly Φ is stationary.

This method of defining the stationary distribution of Φ^o is sometimes inconvenient as it involves defining the limiting procedure. There is also a problem of proving the existence of the limit (2.1). However if Φ^o is a Palm version we can establish the existence and value of the limit (2.1).

In this section Φ^o is a Palm version and its distribution we denote by P^o. We shall denote

$$(2.2) \qquad \lambda^{-1} = \int_F Y_1^{\wedge}(f) P^o(df) \qquad (= EY_1^o) ,$$

i.e. λ is the reciprocal of the mean inter-point distance. Set for any $F \in BF$

$$(2.3) \qquad P(F) = \lambda \sum_{i=0}^{\infty} P^o(Y_1^{\wedge} > i, \tau^{-i}F) \qquad (= \lambda \sum_{i=0}^{\infty} Pr(Y_1^o > i, \tau^i \Phi^o \in F)) .$$

If $\lambda > 0$ then P is a probability measure on F. In what follows λ is always positive.

The following definition turns out to be justified in view of the considerations of this section.

<u>Definition</u> 2.3:

(i) If $\Phi^o = (X^o, N^o)$ is a Palm representation of an r.p.@ p.p. with the distribution P^o then P given by (2.3) is called the *stationary distribution of* Φ^o.

(ii) Any r.p.@ p.p. $\Phi = (X, N)$ with the distribution P is called a *stationary* r.p.@ p.p. *corresponding* to Φ^o.

<u>Proposition</u> 2.3:

For any measurable function $\varphi: F \to R_+$,

$$(2.4) \qquad E\,\varphi(\Phi) = \lambda\, E \sum_{i=0}^{Y_1^o - 1} \varphi(\tau^i \Phi^o) ,$$

where Φ is a stationary r.p.@ p.p. corresponding to Φ^o.

We omitt the standard proof. Notice the following special cases of interest:

(2.5) $$P(X_0^{\wedge} \in F) = \lambda \sum_{i=0}^{\infty} \Pr(Y_1^o > i,\ X_i^o \in F) \ ,$$

(2.6) $$P(Y_1^{\wedge} \geq j) = \lambda \sum_{i=j}^{\infty} \Pr(Y_1^o > i) \ .$$

The following theorem is basic. It is simply an extension of Theorem 1 from Ryll-Nardzewski (1961) onto the case of r.p.@ p.p.'s.

Theorem 2.1:

(i) $\lambda = P(N_0^{\wedge} = 1)$.

(ii) The distribution P defined in (2.3) is stationary.

(iii) If

$$Q(\cdot) = P(\cdot \mid N_0^{\wedge} = 1)$$

then $Q = P^o$.

(iv) If P' is a stationary distribution on $(F, \mathcal{B}F)$ then

$$Q'(\cdot) = P'(\cdot \mid N_0^{\wedge} = 1)$$

is a Palm distribution.

Remark:

From (i) it follows that λ is the *intensity* of the p.p. N ($\Phi = (X,N)$ is a stationary r.p.@ p.p. corresponding to Φ^o).

Proof: The proof follows from (2.3) substituting

$$F = \{f = (d,n) : n_0 = 1\} \ .$$

(ii). For any $F \in \mathcal{B}F$

$$P(\tau^{-1}F) = \lambda \sum_{i=0}^{\infty} P^o(Y_1^{\wedge} > i,\ \tau^{-i-1}F) =$$

$$= \lambda \sum_{i=1}^{\infty} (P^o(Y_1^{\wedge} > i,\ \tau^{-i}F) + P^o(Y_1^{\wedge} = i,\ \tau^{-i}F)) =$$

$$= \lambda \sum_{i=0}^{\infty} P^o(Y_1^{\wedge} > i,\ \tau^{-i}F) + \lambda(\sum_{i=1}^{\infty} P^o(Y_1^{\wedge} = i,\ \tau^{-i}F) - P^o(F)) \ .$$

Now

$$\sum_{i=0}^{\infty} P^o(Y_1^{\wedge} = i,\ \tau^{-i}F) - P^o(F) = P^o(\sigma^{-1}F) - P^o(F) = 0$$

because P^o is a Palm distribution.

(iii). For any $F \in \mathcal{B}F$

$$P(F|N_0^{\wedge}=1) = \frac{P(F|N_0^{\wedge}=1)}{P(N_0^{\wedge}=1)} =$$

$$= \sum_{i=0}^{\infty} P^o(Y_1^{\wedge} > i, \ \tau^{-i}(F \cap \{N_0^{\wedge}=1\})) = P^o(F) \ .$$

(iv). Q' is a Palm distribution iff $Q'\sigma^{-1} = Q'$. Let $F \in \mathcal{B}F$. Then

$$Q'(\sigma^{-1}F) = \frac{P'(\sigma\Phi^{\wedge} \in F, \ N_0^{\wedge}=1)}{P'(N_0^{\wedge}=1)} =$$

$$= \frac{\sum_{k=1}^{\infty} P'(\sigma\Phi^{\wedge} \in F, \ N_0^{\wedge}=1, \ Y_{-1}^{\wedge}=k)}{P(N_0^{\wedge}=1)} =$$

$$= \frac{\sum_{k=1}^{\infty} P'(\Phi^{\wedge} \in F, \ N_0^{\wedge}=1, \ Y_1^{\wedge}=k)}{P'(N_0^{\wedge}=1)} = Q'(F) \ .$$

From Theorem 2.1 it follows that there is a one-to-one correspondence between a stationary distribution P and the respective Palm distribution P^o. Namely with P^o the distribution P is given by (2.3). On the other hand to a stationary P we relate $P^o(\cdot) = P(\cdot|N_0^{\wedge}=1)$. Following Definition 2.3 we call any r.p.@ p.p. $\Phi^o = (X^o, N^o)$ with the distribution P^o a *Palm version corresponding to the stationary* r.p.@ p.p. Φ with the distribution P.

The next theorem shows that the distribution P defined by (2.3) is indeed the looked for stationary distribution of Φ^o. In fact we show a bit more and the requirement (2.1) will be obtained as a corollary.

Theorem 2.2:

Let Φ^o be $\underline{\sigma}$-ergodic defined on $(A, \mathcal{S}A, \Pr)$. If $\varphi: F \to \mathbb{R}$ is a P-integrable function then

$$\text{(2.7)} \qquad \lim_{j\to\infty} \frac{1}{j} \sum_{i=0}^{j-1} \varphi(\tau^i \Phi^o) = E\varphi(\Phi) \ , \qquad \text{a.e.,}$$

where Φ is a stationary r.p.@ p.p. corresponding to Φ^o.
Moreover if G is a Polish space and $\psi: F \to G$ is a measurable mapping then for $\mu = \Pr\,\psi(\Phi^o)^{-1}$ - a.e. points $x \in G$

$$(2.8)\qquad \lim_{j\to\infty} \frac{1}{j} \sum_{i=0}^{j-1} \varphi(\tau^i \Phi^o) = E\varphi(\Phi)\ , \qquad Pr(\cdot\,|\,\psi(\Phi^o) = x)\ -\ \text{a.e.}\ .$$

(ii) Let Φ be a metrically-transitive r.p.@ p.p. defined on a probability space (A, SA, Pr). If $\varphi: F \to R$ is a P^o-integrable function then

$$\lim_{j\to\infty} \frac{1}{j} \sum_{i=0}^{j-1} \varphi(\sigma^i \Phi) = E\varphi(\Phi^o)\ , \qquad \text{a.e.},$$

where Φ^o is a Palm version corresponding to Φ.
Moreover if G is a Polish space and $\psi: F \to G$ is a measurable mapping, then for $\nu = Pr\ \psi(\Phi)^{-1}$ - a.e. points $x \in G$

$$\lim_{j\to\infty} \frac{1}{j} \sum_{i=0}^{j-1} \varphi(\sigma^i \Phi) = E\varphi(\Phi^o)\ , \qquad Pr(\cdot\,|\,\psi(\Phi) = x)\ -\ \text{a.e.}\ .$$

The proof of Theorem 2.2 we preceed by a remark and two lemmas.

Remark:

The second part of the theorem allow us to cover some special cases where instead of dealing with a Palm version we have only the conditional Φ^o under some initial conditions. For example in queueing theory we often suppose that a queue is empty at the begining.

Lemma 2.1:

Let A be an invariant (regarding τ) set on (F, BF). Then $P(A) = 1$ iff $P^o(A) = 1$.

Proof: Since A is an invariant set we have

$$P(A) = \lambda \sum_{i=0}^{\infty} P^o(Y_1^{\wedge} > i,\ \tau^{-i}A) = \lambda \sum_{i=0}^{\infty} P^o(Y_1^{\wedge} > i,\ A).$$

Assuming $P^o(A) = 1$ it follows that $P^o(Y_1^{\wedge} > i,\ A) = P^o(Y_1^{\wedge} > i)$. Hence

$$P(A) = \lambda \sum_{i=0}^{\infty} P^o(Y_1^{\wedge} > i) = 1\ .$$

Conversely if $P(A) = 1$ then

$$P^o(A) = \frac{P(A \cap \{N_0^{\wedge} = 1\})}{P(N_0^{\wedge} = 1)} = 1$$

which completes the proof.

Denote by Φ a stationary r.p.@ p.p. corresponding to Φ^o.

Lemma 2.2:

Φ is ergodic iff Φ^o is σ-ergodic

Proof: For the sake of completeness we set forth Ryll-Nardzewski's proof of his Theorem 3 in the case of p.p.'s. As we noted, there is a one-to-one mapping between stationary probabilities and Palm probabilities on F. Moreover it can be shown that extremal points of the convex set of stationary probabilities are mapped onto extremal points of the convex set of Palm probabilities. Now we need only apply Proposition 1.1 to complete the proof.

Proof of Theorem 2.2: (i) Denote

$$B = \{f \in F: \lim_{j\to\infty} \frac{1}{j} \sum_{i=0}^{j-1} \varphi(\tau^i(f)) = E\varphi(\Phi)\}.$$

From the ergodic theorem, bearing in mind Lemma 2.2, we have $P(B) = 1$. Moreover the set B is invariant. Thus by Lemma 2.1 we have $P^o(B) = 1$. The proof of (2.7) will be complete if we remark that the limit (2.7) exists on the set $(\Phi^o)^{-1}B$ and $Pr((\Phi^o)^{-1}B) = 1$. To prove (2.8) we write

$$1 = P^o(B) = Pr((\Phi^o)^{-1}B) = \int_G Pr((\Phi^o)^{-1}B \mid \psi(\Phi^o) = x)\mu(dx) .$$

Since $0 \le Pr((\Phi^o)^{-1}B \mid \psi(\Phi^o) = x) \le 1$ (we can choose such a version) we obtain

$$1 = Pr((\Phi^o)^{-1}B \mid \psi(\Phi^o) = x) , \qquad \mu - a.e. .$$

(ii) Denote

$$B^o = \{f \in F: \lim_{j\to\infty} \frac{1}{j} \sum_{i=0}^{j-1} \varphi(\sigma^i f) = E\varphi(\Phi^o)\}.$$

From the ergodic theorem $P^o(B^o) = 1$. The rest of the proof is the same as in (i).

Corollary 2.2:

(i) For any $F \in F$

$$\lim_{j\to\infty} \frac{1}{j} \sum_{i=0}^{j-1} Pr(\tau^i \Phi^o \in F) = P(F) . \tag{2.9}$$

(ii) If X_i^o, $i \in Z$ are nonnegative r.v.'s then

$$\lim_{j\to\infty} \frac{1}{j} \sum_{i=0}^{j-1} X_i^o = \int_F \hat{X}_0(f)P(df) , \qquad \text{a.e.} \tag{2.10}$$

Proof: By Theorem 2.2, setting $\phi = 1_F$ we obtain

$$\lim_{j\to\infty} \frac{1}{j} \sum_{i=0}^{j-1} 1_F(\tau^i \Phi^o) = P(F) \quad , \qquad \text{a.e. .}$$

Now (2.9) follows from the Lebesgue bounded convergence theorem. If $\int_F \hat{X}_0(f)P(df) < \infty$ then (2.10) follows from Theorem 2.2. If $\int_F \hat{X}_0(f)P(df) = \infty$ then the assertion can be verified by applying the standard trncation argument.

The results which have been worked out in the section will be used in the next example. There we shall continue the investigations of generalized-regenerative processes which were introduced in Example 2.1.

Example 2.2:

Consider a generalized-regenerative process $\Phi^o = \{X_i^o, N_i^o\}$. It was mentioned in Example 2.1 that Φ^o is a Palm version. We suppose that Φ^o is σ-ergodic and that $ET_0^o < \infty$. From Corollary 2.2 (i), we obtain

$$\lim_{k\to\infty} \frac{1}{k} \sum_{i=0}^{k-1} Pr(\tau^i \Phi^o \in F) = \lambda \sum_{i=0}^{\infty} Pr(Y_1^o > i, \tau^i \Phi^o \in F) ,$$

where $\lambda^{-1} = EY_1^o = ET_0^o$.
In particular

$$\lim_{k\to\infty} \frac{1}{k} \sum_{j=0}^{k-1} Pr(X_j^o \in E) = \lambda \sum_{i=0}^{\infty} Pr(Y_1^o > i, X_i^o \in E) .$$

Assuming more $X_i^o \geq 0$, $i \in Z$, from Corollary 2.2 (ii)

$$\lim_{k\to\infty} \frac{1}{k} \sum_{i=0}^{k-1} X_i^o = EX_0 , \qquad \text{a.e..} \tag{2.11}$$

From (2.4) it follows an identity of Wald's type, namely

$$EX_0 = \frac{E \sum_{i=1}^{Y_1^o} Z_{0i}^o}{EY_1^o} . \tag{2.12}$$

§ 3. First and second type relations.

In this section $\Phi^o = \{X_i^o, N_i^o\}$ denotes an r.p.@ p.p. which is a Palm version. Let P^o denote the distribution of Φ^o and P denote the stationary distribution of Φ^o. We are looking for relations between

$$(2.13) \qquad v(E) = P(\hat{X}_0 \in E) \ ,$$

$$(2.14) \qquad v^+(E) = \Pr(X_0^o \in E) = P^o(\hat{X}_0 \in E) \ ,$$

$$(2.15) \qquad v^-(E) = \Pr(X_{-1}^o \in E) = P^o(\hat{X}_{-1} \in E) \ , \qquad E \in \mathcal{B}E \ .$$

In the theory of applied random processes we can sometimes obtain directly v^+ and v^-. Then using the looked for relations we would be able to find v. The looked for relations will be expressed in terms of some characteristics of the r.p.@ p.p. Φ^o which have intuitive meanings.

Let $\{p_i(x,E),\ x\in E,\ E\in\mathcal{B}E,\ i\in\mathbb{N}_0\}$, be a family of stochastic kernels on $(E,\mathcal{B}E)$ such that

$$(2.16) \qquad \Pr(X_i^o \in E \mid X_0^o, Y_1^o > i) = p_i(X_0^o, E) \ , \quad \text{a.e.}.$$

The assumption that E is Polish insures that we can choose the family $\{p_i(x,E)\}$ such that $p_0(x,E) = 1_E(x)$. The family $\{p_i(x,E)\}$ describes the behaviour of the r.p. $\{X_i^o\}$ within intervals between consecutive points of the p.p. $\{N_i^o\}$.

Let $\{f_j^x,\ x\in E\ ,\ j\in\mathbb{N}\}$ be a family of $\mathcal{B}E$-measurable functions such that

(a) for any $j\in\mathbb{N}$

$$(2.17) \qquad \Pr(Y_1^o = j \mid X_0^o) = f_j^{X_0^o} \ , \qquad \text{a.e.} ,$$

(b) $\sum_{j=1}^{\infty} f_j^x = 1 \ , \quad f_j^x \ge 0, \quad x\in E, \quad j\in\mathbb{N}.$

The family $\{f_j^x\}$ describes the distance to the nearest point to the right of zero if X_0^o is given.

Let $\{q(x,E),\ x\in E,\ E\in\mathcal{B}E\}$ be a stochastic kernel such that

$$\Pr(X_0^o \in E \mid X_{-1}^o) = q(X_{-1}^o, E) \ , \qquad \text{a.e.}.$$

The stochastic kernel $\{q(x,E)\}$ describes the transitions of the process at instants from the p.p. N^{o}.

Recall that $\lambda^{-1} = EY_0^{o}$.

Theorem 2.3 (the first type relation):

$$v(E) = \lambda \sum_{i=0}^{\infty} \int_E v^{+}(dx)\Big(\sum_{j>i} f_j^{x}\Big)p_i(x,E) \ , \qquad E \in \mathcal{BE}.$$

Proof: Let $E \in \mathcal{BE}$. Then by (2.5) we have

$$v(E) = \lambda \sum_{i=0}^{\infty} \Pr(Y_1^{o} > i,\ X_i^{o} \in E) \ .$$

Now we transform

$$\lambda \sum_{i=0}^{\infty} \Pr(Y_1^{o} > i,\ X_i^{o} \in E) = \lambda \sum_{i=0}^{\infty} E\, 1_{(i,\infty)}(Y_1^{o})1_E(X_i^{o})$$

$$= \lambda \sum_{i=0}^{\infty} E(E\, 1_{(i,\infty)}(Y_1^{o})1_E(X_i^{o})\,|\,X_0^{o}, Y_1^{o} > i)$$

$$= \lambda \sum_{i=0}^{\infty} E(1_{(i,\infty)}(Y_1^{o})E(1_E(X_i^{o})\,|\,X_0^{o}, Y_1^{o} > i))$$

$$= \lambda \sum_{i=0}^{\infty} E\, 1_{(i,\infty)}(Y_1^{o})p_i(X_0^{o},E)$$

$$= \lambda \sum_{i=0}^{\infty} E(E(1_{(i,\infty)}(Y_1^{o})p_i(X_0^{o},E)\,|\,X_0^{o}))$$

$$= \lambda \sum_{i=0}^{\infty} E(p_i(X_0^{o},E)E(1_{(i,\infty)}(Y_1^{o})\,|\,X_0^{o}))$$

$$= \lambda \sum_{i=0}^{\infty} \int_E v^{+}(dx)\Big(\sum_{j>i} f_j^{x}\Big)p_i(x,E)$$

which completes the proof.

Before we state the second type relation we find a representation of v^{-}.

Proposition 2.4:

If for any $E \in \mathcal{BE}$, $k=1,2,\ldots$

$$\text{(2.18)} \qquad \Pr(X_{k-1}^{o} \in E\,|\,Y_1^{o} = k,\ X_0^{o})$$

$$= \Pr(X_{k-1}^{o} \in E\,|\,Y_1^{o} > k-1, X_0^{o}) \qquad (= p_{k-1}(X_0^{o},E)) \qquad \text{a.e.}$$

then

$$(2.19)\qquad v^-(E) = \sum_{i=1}^{\infty} \int_E v^+(dx) f_i^x p_{i-1}(x,E) , \qquad E \in \mathcal{B}E .$$

Proof: Since Φ^o is a Palm version we have

$$\Pr(X^o_{-1} \in E) = \Pr(X^o_{Y^o_1 - 1} \in E) .$$

Now

$$\Pr(X^o_{Y^o_1-1} \in E) = \sum_{k=1}^{\infty} E(\Pr(X^o_{Y^o_1-1} \in E \mid Y^o_1 = k,\ X^o_0)\Pr(Y^o_1 = k \mid X^o_0))$$

$$= \sum_{k=1}^{\infty} E \Pr(X^o_{k-1} \in E \mid Y^o_1 = k,\ X^o_0) f_k^{X^o_0}$$

which by (2.18) is equal to

$$\sum_{k=1}^{\infty} \int_E v^+(dx) \Pr(X^o_{k-1} \in E \mid Y^o_1 > k-1,\ X^o_0 = x) f_k^x .$$

We now state a theorem in which we give the second type relation. Recall that a family of transition functions $\{\pi_i(x,E)\}$ is Markovian if

$$(2.20)\qquad \pi_{i+j}(x,E) = \int_E \pi_i(x,dy)\pi_j(y,E) , \qquad i,j \in N_o,\quad x \in E,\quad E \in \mathcal{B}E.$$

Theorem 2.4 (the second type relation):

If the family of transition functions $\{p_i(x,E)\}$ is Markovian then

$$(2.21)\qquad \int_E v^+(dx)p_1(x,E) - v(E) = \lambda(v^-(E) - v^+(E)) .$$

Proof: Using the first type relation and assuming that the family of transition functions $\{p_i(x,E)\}$ is Markovian we have

$$\int_E v(dx)p_1(x,E) - v(E) =$$

$$= \lambda\Big(\sum_{i=0}^{\infty} \int_E v^+(dx)\Big(\sum_{j>i} f_j^x\Big)p_{i+1}(x,E) - \sum_{i=0}^{\infty} \int_E v^+(dx)\Big(\sum_{j>i} f_j^x\Big)p_i(x,E)\Big)$$

$$= \lambda(v^-(E) - v^+(E))$$

which completes the proof.

Note that the operator, on the space of finite measures on $(E,\mathcal{B}E)$ defined by

$$\mathcal{U}_p:\ m(\cdot) \to \int_E m(dx)p_1(x,\cdot) - m(\cdot)$$

is a discrete analogue of an infinitesimal operator.

Corollary 2.4:

$$\mathcal{U}_p v(\cdot) = \lambda(v^-(\cdot) - \int_E v^-(dx)q(x,\cdot)) . \tag{2.22}$$

Example 2.3:

Consider a discrete time GI/GI/1 queue with the generic sequence $\{T_i^o,S_i^o\}$. Indeed $T_i^o,S_i^o \in N$, $i \in Z$. We assume the stability condition that

$$\frac{ES_0^o}{ET_0^o} < 1 . \tag{2.23}$$

We look for relationships between the stationary d.f. of the virtual waiting time and the stationary d.f. of the actual waiting time in the queue. Consider an a.w.t.p. $\{W_i^o\}$. Because of (2.23) we can suppose that $\{W_i^o\}$ is a stationary sequence of finite r.v.'s. Instants of arrival of customers to the system are

$$Y_j^o = \begin{cases} \sum_{i=1}^{j-1} T_i^o , & j=1,2,\dots \\ 0 , & j=0 \\ -\sum_{i=-1}^{j} T_i^o , & j=-1,-2,\dots \end{cases}$$

Then we define the v.w.t.p. $\{V_j^o\}$ by

$$V_j^o = [W_i^o + S_i^o + Y_i^o - j]^+ , \qquad Y_i^o \le j < Y_{i+1}^o .$$

The p.p. N^o has points in $\{Y_i^o\}$, i.e. $N_j^o = 1$ if and only if for some $i \in Z$ we have $Y_i^o = j$. Bearing in mind that $\{W_i^o,T_i^o,S_i^o\}$ is stationary (see Loynes' lemma) we can prove that $\Phi^o = (V^o,N^o)$ is a Palm version. We can also find that

$$f_j^x = Pr(T_0^o = j) ,$$

$$p_i(j,\{k\}) = 1_{\{k\}}([j-i]^+) ,$$

$$q(i,\{j\}) = Pr(S_0^o = j-i) ,$$

$$\lambda^{-1} = ET_0^o .$$

Moreover it can be shown that (2.18) is fulfilled. Note the important fact that

$$V^o_{Y_i^o - 1} = W_i^o , \qquad i \in Z .$$

Thus the imbedded chain in $\{V_i^o\}$ at points just before points of N is an a.w.t.p.. Thus v^- is the stationary d.f. of the actual waiting time. Denote by $\Phi = (V,N)$ a stationary r.p.@ p.p. corresponding to Φ^o and the distribution of Φ by P. The searched stationary d.f. of the virtual waiting time is $v(E) = P(V_0^\wedge \in E)$. Using now Theorem 2.3 and 2.4 we have

$$v(\{k\}) = \begin{cases} \lambda \sum_{i=0}^{\infty} \left(\sum_{j>i} f_j\right) \sum_{l=0}^{k+i} v^-(\{l\}) b_{k+i-l} , & k=1,2,\ldots \\ \lambda \sum_{i=0}^{\infty} \left(\sum_{j>i} f_j\right) \sum_{j=0}^{i} \sum_{l=0}^{j} v^-(\{l\}) b_{j-l} , & k = 0 \end{cases}$$

and

$$v(\{k+1\}) - v(\{k\}) = \lambda\left(v^-(\{k\}) - \sum_{j=0}^{k} v^-(\{j\}) b_{k-j}\right) , \quad k=0,1,\ldots ,$$

where

$$b_j = Pr(S_0^o = j) .$$

Thus $\{v\}$ and $\{v^-\}$ are such that they fulfill the system of equations. I conjecture that the system has a unique solution. In such a case we would have a way of finding the stationary d.f. of the virtual waiting time and the stationary d.f. of the actual waiting time.

We finish up the section by outlining a method of solving the Poisson type equation

$$\mathcal{U}_p v = -m$$

with an unknown v. If v^-, λ, $\{p_i(x,E)\}$, $\{q(x,E)\}$ are known then v is a solution of the equation (2.22) which is indeed of that type. Let M be the space of all signed measures with a finite variation on (E,BE). The set M endowed with the variation norm $\|w\|$ is a Banach space. Denote by S the set of invariant measures from M with respect to $\{p_1(x,E)\}$, i.e.

$$S = \{w \in M: \int_E w(dx)p_1(x,\cdot) = w(\cdot)\} .$$

Proposition 2.5:

If

$$(2.24) \qquad \sum_{k=0}^{\infty} \|\int_E m(dx)p_k(x,\cdot)\| < \infty$$

then the only solution of the equation

$$(2.25) \qquad U_p v = -m$$

is of the form

$$v(\cdot) = \sum_{k=0}^{\infty} \int_E m(dx)p_k(x,\cdot) + z(\cdot) , \quad z \in S .$$

Proof: From (2.24) it follows that

$$\lim_{l\to\infty} \sum_{k=0}^{l} \int_E m(dx)p_k(x,\cdot) = \sum_{k=0}^{\infty} \int_E m(dx)p_k(x,\cdot),$$

where the limit is taken with respect to the norm $\| \ \|$. Thus

$$\lim_{l\to\infty} U_p \sum_{k=0}^{l} \int_E m(dx)p_k(x,\cdot) = U_p \sum_{k=0}^{\infty} \int_E m(dx)p_k(x,\cdot) .$$

However

$$U_p \sum_{k=0}^{l} \int_E m(dx)p_k(x,\cdot) = -m(\cdot) - \int_E m(dx)p_{l+1}(x,\cdot) .$$

Hence taking advantage of the fact that (by (2.24))

$$\lim_{k\to\infty} \|\int_E m(dx)p_k(x,\cdot)\| = 0$$

we have

$$\mathcal{U}_p \sum_{k=0}^{\infty} \int_E m(dx)p_k(x,\cdot) = -m(\cdot) .$$

This shows that $\sum_{k=0}^{\infty} \int_E m(dx)p_k(x,\cdot) = -m(\cdot)$ is a solution of (2.25). Suppose now that there exists $w \in M$ which is also a solution of (2.25) and is different from $\sum_{k=0}^{\infty} \int_E m(dx)p_k(x,\cdot)$. Then we get

$$\mathcal{U}_p(w(\cdot) - \sum_{k=0}^{\infty} \int_E m(dx)p_k(x,\cdot)) = \mathcal{U}_p w(\cdot) = 0 .$$

Thus $w \in S$ which completes the proof.

§ 4. Applications in queueing theory .

In this section we study three problems.

(i) The stability of work-conserving normal queues.

(ii) Is the a.w.t.p. in a G/G/1; FIFO queue a generalized-regenerative process? Crane & Iglehart noticed that in a case of GI/GI/1 queues this is true.

(iii) A representation of the a.w.t.p. in a $GI^{GI}/GI/1$; FIFO queue in terms of the a.w.t.p. in a G/GI/1 queue.

Notice that the above three problems are not from the theory of discrete time queueing systems. Such a theory is not very interesting as it is similar to the theory of continuous time queueing systems. All the above mentioned problems have no counterparts in continuous time.

Let τ denote the shift transformation on E^Z defined by

$$\tau(\{x_i, i \in Z\}) = \{x_{i+1}, i \in Z\}$$

Recall that a sequence $X = \{X_i\}$ is stable if there exists an honest stationary distribution P on (E^Z, BE^Z) such that

$$\lim_{j\to\infty} \frac{1}{j} \sum_{i=0}^{j-1} \Pr(\tau^i X \in \cdot) = P(\cdot)$$

and is strongly stable if

$$\lim_{i\to\infty} \Pr(\tau^i X \in \cdot) = P(\cdot).$$

Now we begin with the problem (i). Consider a G/G/1 queue with the work-conserving, normal discipline. Let $\{T_i^o, S_i^o\}$ be a generic sequence of the queue. We suppose that the r.e.'s used herein are defined on a probability space (A, SA, Pr).

Theorem 2.5:

If

$$\frac{ES_0^o}{ET_0^o} < 1 \tag{2.26}$$

then the queue is strongly stable.

Proof: Let $\tilde{W} = \{\tilde{W}_i\}$ be an a.w.t.p. in the queue. We construct a stationary a.w.t.p. $W = \{W_i\}$ and then find that conditions of Lemma 1.2 are fulfilled. This would mean that the constructed r.p. W is a stationary a.w.t.p. corresponding to $\tilde{W}$. Let

$$\tilde{Y}_1 = \min\{i > 0 : \tilde{W}_i = 0\}$$

and define $W^* = \{W_i^*\}$ a coupled with $\tilde{W}$ a.w.t.p. in a G/G/1; FIFO queue by

$$W_{i+1}^* = \begin{cases} 0, & i < \tilde{Y}_1, \\ \max(W_i^* + S_i^o - T_i^o), & i \geq \tilde{Y}_1. \end{cases}$$

Define $\tilde{N}$, N^*, associated with $\tilde{W}$ and W^* p.p.'s, by

$$\tilde{N} = \{\delta_0(\tilde{W}_i)\},$$

$$N^* = \{\delta_0(W_i^*)\}.$$

The stability condition (2.26) ensures the existence of a finite metrically-transitive sequence $M = \{M_i\}$ on (A, SA, Pr) fulfilling

$$M_{i+1} = \max(0, M_i + S_i^o - T_i^o), \quad i \in Z.$$

Set

$$N = \{\delta_0(M_i)\}.$$

The (M,N) is a stationary r.p.@ p.p. corresponding to (W^*, N^*).

Moreover there exists a nonnegative integer-valued r.v. Z such that

$$(2.27) \qquad N_{Z+i} = N^*_{Z+i}\,, \qquad i=0,1,\ldots\,.$$

This can be shown by an argument similar to the one used by Loynes (1962), Section 2.32. Due to the assumption that the queue is work-conserving we obtain

$$(2.28) \qquad \tilde{N} = N^*\,.$$

From Loynes' lemma the sequence $\{(T_i^o, S_i^o), N_i\}$ is stationary. Let $W = \{W_i\}$ be the a.w.t.p. in the queue defined by the use of function φ (from the definition of a normal discipline) applied to $\{T_i^o, S_i^o\}$, where the consecutive busy periods are determined by points of N. We shall show that $\{W_i\}$ is the looked for stationary a.w.t.p. in the queue. First we note that due to Proposition 1.2 the r.p. $\mathbf{W}$ is stationary. Clearly

$$N = \{\delta_0(W_i)\}\,.$$

Thus by (2.27) and (2.28) we have that

$$\tilde{N}_{Z+i} = N_{Z+i}\,, \qquad i=0,1,\ldots$$

and using the assumption of normality of the discipline

$$W_{Z+i} = \tilde{W}_{Z+i}\,, \qquad i=0,1,\ldots\,.$$

This, by Lemma 1.2 completes the proof.

Remark:

We proved in fact that $\{W_i, T_i^o, S_i^o, K_i^o\}$ is stationary. This will be used in Chapter 5.

Now consider a G/G/1 queue with a work-conserving, normal discipline determined by a generic sequence $\{T_i^o, S_i^o, K_i^o\}$. Assume that

(i) $\{T_i^o, S_i^o, K_i^o\}$ is metrically-transitive,

(ii) $\dfrac{ES_0^o}{ET_0^o} < 1\,,$

(iii) $K_i^o \in \{1,\ldots,l\}\,.$

We read that the i-th customer arriving in the system is from the K_i^o-th class and obtains the service S_i^o. Let $\{\tilde{W}_j\}$ be an a.w.t.p. in the queue. The a.w.t.p. of customers from the i-th class $i \in \{1,\dots,l\}$ is defined as follows. From the sequence $\{\tilde{W}_j\}$ we cancel out all $\tilde{W}_j$ such that $K_j^o \neq i$. Denote the ensuing sequence by $\{\tilde{W}_{ij}, j \in Z\}$.

Corollary 2.5:

For any $i \in \{1,\dots,l\}$ the sequence $\{\tilde{W}_{ij}, j \in Z\}$ is strongly stable.

Proof: Let $\{W_j\}$ be the stationary a.w.t.p. in the queue. Such a finite sequence exists by Theorem 2.5. From the sequence $\{W_j\}$ we cancel out all W_j such that $K_j^o \neq i$ and the ensuing sequence we denote by $\{W_{ij}, j \in Z\}$. Indeed $\{W_{ij}, j \in Z\}$ is a stationary sequence. We know that for some r.v. Z

$$\tilde{W}_{Z+j} = W_{Z+j}, \qquad j=0,1,\dots .$$

Hence for some r.v. Z'

$$\tilde{W}_{i,Z'+j} = W_{i,Z'+j}, \qquad j=0,1,\dots$$

and the assertion of the corollary follows from Lemma 1.2.

Problem (ii). In this problem it is more convenient to consider r.p.@ p.p.'s with time running through N_0 instead of Z. Thus before we state the details of the problem we notice a possible way of modifying the theory. An r.p.@ p.p. $\Phi^o = \{(X_i^o,N_i^o), i \in N_0\}$ is said to be a Palm version if there exists a Palm version $\{(X_i',N_i'), i \in Z\}$ such that $\{(X_i',N_i'), i \in N_0\}$ and $\{(X_i^o,N_i^o), i \in N_0\}$ are identically distributed. Thus if P^o denotes the distribution of Φ^o then Φ^o is a Palm version if and only if $P^o\sigma^{-1} = P^o$, where

$$\sigma\{(x_i,n_i), i \in N_0\} = \{(x_{i+y_1},n_{i+y_1}), i \in N_0\}$$

and y_1 is the coordinate of the first point of n on N. Similarly other concepts defined for r.p.@ p.p. may be modified.

Let $\{T_i^o,S_i^o\}$ be a metrically-transitive sequence. Consider the r.p.@ p.p. $\{(W_i^*,N_i^*), i \in N_0\}$ defined by

$$W_0^* = 0 ,$$

$$W^*_{i+1} = \max(0, W^*_i + S^o_i - T^o_i) , \qquad i=0,1,\ldots$$

and

$$N^*_i = \delta_0(W^*_i) , \qquad i=0,1,\ldots .$$

The r.p. $\{W^*_i, i \in N_0\}$ is an a.w.t.p. in a G/G/1; FIFO queue. It is known (see e.g. Crane & Iglehart (1974)) that in GI/GI/1; FIFO queues the r.p.@ p.p. $\{(W^*_i, N^*_i), i \in N_0\}$ is regenerative. Unfortunately, this process need not even be generalized-regenerative as the following example shows. Let $A = \{\underline{a}_1, \underline{a}_2\}$ and $Pr(\underline{a}_1) = Pr(\underline{a}_2) = \frac{1}{2}$, where

$$\underline{a}_i = (a_{i0}, a_{i1}, \ldots) , \qquad i=1,2 .$$

Set

$$a_{ij} = \begin{cases} (2,b) , & \text{if } i+j \text{ is odd,} \\ (1,b) , & \text{if } i+j \text{ is even.} \end{cases}$$

The generic sequence is defined by

$$(T^o_j(\underline{a}_i), S^o_j(\underline{a}_i)) = a_{ij} , \qquad i=1,2, \qquad j=0,1,\ldots .$$

Clearly $\{T^o_i, S^o_i\}$ is metrically-transitive. The traffic intensity is

$$\rho = \frac{ES^o_0}{ET^o_0} = \frac{2}{3} b .$$

Thus if $1 < b < \frac{3}{2}$ then the sequence $\{W^*_i, i \in N_0\}$ is stable. Moreover $\{W^*_i, i \in N_0\}$ assumes values

$$(0,0,b-1,0,b-1,\ldots)$$

and

$$(0,b-1,0,b-1,\ldots)$$

with probability $\frac{1}{2}$. However $\{W^*_{Y^*_1+i}, i \in N_0\}$ assumes with probability 1 the value

$$(0,b-1,0,\ldots) ,$$

where Y^*_1 denotes the first positive instant i such that $W^*_i = 0$.

Thus $\{(W_i^*, N_i^*),\ i \in N_0\}$, with the distribution P^*, may not be generalized-regenerative because it would require that $P^* \sigma^{-1} = P^*$. Nevertheless a question emerges about a generalized-regenerative representation of $\{(W_i^*, N_i^*),\ i \in N_0\}$ in G/G/1 queues. Such a representation can be found. Let P be the stationary distribution of $\{(W_i^*, N_i^*),\ i \in N_0\}$. Then any r.p. $\{(W_i^o, N_i^o),\ i \in N_0\}$ with the distribution P^o is the looked for generalized-regenerative representation (P^o denotes the Palm distribution corresponding to P). From (2.12) we have

$$EW_0 = \frac{E \sum_{i=0}^{Y_1^o - 1} W_i^o}{EY_1^o} ,$$

where $\{W_i,\ i \in N_0\}$ is a stationary r.p. corresponding to $\{W_i^o,\ i \in N_0\}$.

(iii) We will discuss briefly the third problem. It deals with a $GI^{GI}/GI/1$; FIFO queue, namely a queue at which customers arrive in batches. The sizes of the batches are i.i.d.r.v.'s and they are independent of the input and service. The generic sequence of such a queue is

$$\{T_i^o,\ (D_i^o,\ S_{i1}^o,\ \ldots,\ S_{iD_i^o}^o)\} .$$

Here

T_i^o denotes the inter-arrival time between the i-th and (i+1)-st customer,

D_i^o denotes the size of the i-th batch,

S_{ij}^o denotes the service time of the j-th customer in the i-th batch.

We assume that all r.e.'s considered herein are defined on a common probability space (A, SA, Pr). All components of the generic sequence, namely $\{T_j^o\}$, $\{D_j^o\}$, $\{S_{ij}^o,\ j \in Z\}$ are independent and consists of i.i.d. r.v.'s. Assume a stability condition

$$\frac{ES_{01}^o}{ED_0^o\ ET_0^o} < 1 . \tag{2.29}$$

Let $\tilde{W}^{(1)}$ be an a.w.t.p. of the first customers in batches. It fulfills the following recursive relationship

$$(2.30)\qquad \tilde{W}^{(1)}_{i+1} = \begin{cases} 0 , & i=-1,-2,\dots \\ \max(0,\ \tilde{W}^{(1)}_{i} + \sum_{j=1}^{D^o_i} S^o_{ij} - T^o_i) , & i=0,1,\dots . \end{cases}$$

Notice that nothing is assumed on $W^{(1)}_0$. Because of (2.29) there exists a sequence $M^{(1)}$ of finite r.v.'s fulfilling

$$M^{(1)}_i = \max(0,\ M^{(1)}_i + \sum_{j=1}^{\tilde{D}^o_i} S^o_{ij} - T^o_i) , \qquad i \in Z ,$$

and such that

$$(M^{(1)}_i,\ T^o_i,\ D^o_i,\ \{S^o_{ij},\ j=1,\dots\})$$

is stationary. Let $\{\tilde{V}_i\}$ denote an a.w.t.p. in the queue. We look for a relationship between the stationary d.f. of $\tilde{W}^{(1)}$ and the stationary d.f. of $\tilde{V}$. We also find that the stationary distribution of $\tilde{V}$ is the stationary d.f. of the a.w.t.p. in a conjugate G/GI/1; FIFO queue. In a special case, where the sizes of the batches are geometrically distributed, a conjugate queue is of type GI/GI/1; FIFO. Define for $j \in Z$

$$Y^o_j = \begin{cases} \sum_{i=0}^{j-1} D^o_i , & j=1,2,\dots , \\ 0 , & j=0 , \\ -\sum_{i=-1}^{j} D^o_i , & j=-1,-2,\dots , \end{cases}$$

$$\nu(j) = \max\{i:\ Y^o_i \le j\} ,$$

$$V^o_j = \begin{cases} M^{(1)}_{\nu(j)} , & j = Y^o_{\nu(j)} \\ M^{(1)}_{\nu(j)} + \sum_{i=1}^{j-Y^o_{\nu(j)}} S^o_{\nu(j),i} , & j > Y^o_{\nu(j)} , \end{cases}$$

$$N^o_j = \begin{cases} 0 , & j \notin \{Y^o_i\} \\ 1 , & j \in \{Y^o_i\} . \end{cases}$$

The r.p.@ p.p. $\{V_j^o, N_j^o\}$ is a Palm version. Let $\{V_j, N_j\}$ be a stationary r.p.@ p.p. corresponding to $\{V_j^o, N_j^o\}$. To find the conjugate G/GI/1; FIFO queue set

$$T_j^* = \begin{cases} 0 , & N_j^o = 0 , \\ T_{\nu(j)}^o , & N_j^o = 1 , \end{cases}$$

$$S_j^* = S^o_{\nu(j), j+1-Y^o_{\nu(j)}} , \qquad j \in Z .$$

The r.p.@ p.p. $\{(T_j^*, S_j^*), N_j^o\}$ is a Palm version. Let $\{(T_i, S_i), N_i\}$ be a stationary r.p.@ p.p. corresponding to $\{(T_j^*, S_j^*), N_j^o\}$. Let $\{M_i\}$ be the stationary a.w.t.p. in a G/GI/1; FIFO queue with the generic sequence $\{T_i, S_i\}$. This queue is a looked for G/GI/1; FIFO representation of the a.w.t.p. in the $GI^{GI}/GI/1$; FIFO queue.

Properties of the sequences $\{S_i\}$, $\{T_i\}$ and $\{M_i\}$ are collected in the next proposition. They are stated without proofs. Proofs require lot of formal but routine checkings.

Proposition 2.6:

(i) $\{S_i\}$ is a stationary independent sequence and r.v.'s S_0, S_{01}^o are identically distributed.

(ii)

$$ET_0 = \frac{ET_0^o}{ED_0^o} .$$

(iii) The r.p.'s $\{V_i\}$ and $\{M_i\}$ are identically distributed.

(iv) If D_i^o is a stationary independent sequence of geometrically distributed r.v.'s with the mean $1/(1-a)$ then the sequence $\{T_i\}$ consists of i.i.d.r.v.'s with the common d.f. $1-a+a\Pr(T_0^o \le x)$, $x \ge 0$.

Theorem 2.6:

(i) Under condition (2.29) the r.p. $\tilde{V}$ is strongly stable.

(ii) For $A \in \mathcal{B}R_+$

$$\Pr(V_0 \in A) = \frac{1}{ED_0^o} \int_0^\infty \Pr(M_0^{(1)} \in dx) \sum_{i=0}^\infty \Pr(D_0^o > i)\, B^{*i}(A-x) , \tag{2.31}$$

where

$$B(A) = Pr(S^o_{01} \subset A) .$$

In particular

$$(2.32) \qquad EV_0 = EM_0^{(1)} + \frac{ES^o_{01}}{2}\left(\frac{E(D^o_0)^2}{E\,D^o_0} - 1\right).$$

Proof: (i). There exists an r.v. Z such that

$$\tilde{W}^{(1)}_{Z+j} = M^{(1)}_{Z+j} , \qquad j=0,1,\ldots .$$

This can be shown by an argument similar to the one used by Loynes (1962), Section 2.32. Therefore there exists an r.v. Z' such that

$$\tilde{V}_{Z'+j} = M_{Z'+j} , \qquad j=0,1,\ldots .$$

Thus by Lemma 1.2 the proof of (i) follows.
(ii) To prove (2.31) we apply Theorem 2.3 to the r.p.@ p.p. $\{V^o_i, N^o_i\}$. It is sufficient to note that

$$V^o_{Y^o_i} = M^{(1)}_i \qquad i=\ldots,-1,0,1,\ldots .$$

Corollary 2.6:
If the sizes of batches are geometrically distributed with the expected size of a batch equal to $\frac{1}{1-a}$ then

$$Pr(V_0 \in A) = \sum_{i=0}^{\infty} (1-a)a^i \int_0^{\infty} Pr(M_0^{(1)} \in dx)B^{*i}(A-x)$$

and

$$(2.33) \qquad EV_0 = EM_0^{(1)} + \frac{aES^o_{01}}{1-a} .$$

Now we use the formula (2.33) to get a bound for the mean stationary actual waiting time in a queueing system with batch arrivals. Denote

$$\gamma = \frac{E^2T^o_0}{E[T^o_0]^2} .$$

Proposition 2.7:

Suppose that there exists $\beta > 0$ such that

$$(2.34) \qquad \int_x^\infty \Pr(S_0^o > t)dt \le \frac{1}{\beta} e^{-\beta x}, \qquad x \ge 0$$

and there exists $0 \le a < 1$ such that

$$(2.35) \qquad \sum_{i=j}^\infty \Pr(D_0^o \ge i) \le a^{j-1}, \qquad j=1,2,\dots .$$

Moreover we assume that

$$(2.36) \qquad 1 < m_1(1-a)\beta .$$

Then

$$(2.37) \qquad EW^{(1)} \le \frac{1-\gamma+\gamma l}{\beta\gamma(1-a)(1-l)},$$

where $0 < l < 1$ is the only root of the equation

$$l = e^{(-r+rl)}, \qquad r = (1-a)\beta ET_0^o .$$

Proof: Denote by $\{\overline{W}_i^{(1)}\}$ the stationary a.w.t.p. of the first customer in a batch and by $\{\overline{V}_i\}$ the stationary a.w.t.p. in a $GI^{GI}/GI/1$ queue having exponentially distributed service time (with the parameter β) and geometrically distributed sizes of batches (with the parameter a). Such processes exist because of the stability condition (2.36). Using Theorem 3.1 of Rolski (1976), due to the assumptions (2.34), (2.35) we have

$$(2.38) \qquad EW_0^{(1)} \le E\overline{W}_0^{(1)} .$$

By Corollary 2.6

$$(2.39) \qquad E\overline{W}_0^{(1)} = E\overline{V}_0 - \frac{a}{\beta(1-a)} .$$

Proposition 2.6 asserts that $E\overline{V}_0$ is the mean stationary waiting time in a GI/M/1 queue with the inter-arrival time distribution $1-a+a\Pr(T_0^o \le x)$, $x \ge 0$ and the exponential service time with the parameter β. Now using the Corollary after Theorem 4.1 of Rolski (1976), we obtain

$$E\overline{V}_0 \leq \frac{1-(1-a)\gamma+(1-a)\gamma l}{(1-a)\beta\gamma(1-l)}.$$

Hence by (2.38) and (2.39) we get (2.37).

Remark:

The inequality (2.34) means that the service time distribution is less, regarding the convex ordering, than the exponential d.f. with the parameter β. Similarly (2.35) means that the d.f. of D_0^o is less than the geometrical d.f. with the mean $1/(1-a)$. Details about the convex ordering and conditions for (2.34) and (2.35) can be found for example in Stoyan (1977a) or Rolski (1976), (1977b).

Notes

Most of the concepts and theorems of § 1, 2 are extentions of Ryll-Nardzewski's (1961) ones for the case of r.p.@ p.p.'s. In a special case (for piecewise Markow processes) the first and second kind relations were derived by Kopocińska (1977a). The formula (2.33) was obtained by Daley & Trengove (1977).

CHAPTER 3. CONTINUOUS TIME r.p.@ m.p.p.

§ 1. Random process associated with marked point process.

In continuous-time theory we use the same notations as we did in the discrete-case. This chapter starts with the definition of an r.p.@ m.p.p. . One component of an r.p.@ m.p.p. is a p.p. and we define it now. Since points are not always homogeneous we shall deal with so called m.p.p.'s. Let K be a Polish space. The space K is called a *space of marks* and elements of it are called *marks*. Define

$$N_K = N(R \times K) ,$$

where $N(R \times K)$ were introduced in Section 1.2. We write simply N if K consists of a single point.

Definition 3.1:

A measurable mapping

$$N: (A, SA, Pr) \to (N_K, BN_K)$$

such that

$$\text{(3.1)} \qquad Pr(N(\{t\} \times K) \le 1, \ t \in R) = 1$$

is said to be a *marked point process* (m.p.p.). If K consists of a single point then N is called simply a *point process* (p.p.).

Remark:

To justify (3.1) we should verify that

$$A = \bigcap_{t \in R} \{N(\{t\} \times K) \le 1\} \in SA .$$

To prove it recall (see Kallenberg (1976), Lemma 2.3) that there exists a sequence of r.v.'s $\{Y_i\}$ and a sequence of r.e.'s $\{K_i\}$;

$$K_i : (A, SA, \mathrm{Pr}) \to (K, BK) , \qquad i \in Z$$

such that

$$N = \sum_i \delta_{(Y_i, K_i)} .$$

Hence

$$A = \bigcap_{i \neq j} \{Y_i \neq Y_j\} .$$

We read that the m.p.p. has a point at Y_i and that the mark K_i is attached to the point at Y_i, $i \in Z$.
This is justified by (3.1). In a case where K consists of a single point the condition (3.1) means that N is *without multiple points*.
Let E be a Polish space and

$$X : (A, SA, \mathrm{Pr}) \to (\mathcal{D}(R,E), B\mathcal{D}(R,E)) .$$

<u>Definition</u> 3.2:

$\Phi = (X,N)$ is said to be a *random process associated with a marked point process* (r.p.@ m.p.p.).

The r.p.@ m.p.p. Φ assumes values in

$$F = \mathcal{D}(R,E) \times N_K$$

which is clearly a Polish space.

A frequently encountered r.p.@ m.p.p. is either in the form of a mapping

$$N_K \ni n \to (\varphi(n), n) \in F ,$$

where $\varphi : N_K \to \mathcal{D}(R,E)$ is measurable, or in the form of a mapping

$$L \ni d \to (d, \psi(d)) \in F ,$$

where $\psi : \mathcal{D}(R,E) \to N_K$ is measurable. For example $\psi(X)$ may be a family of instants of jumps in a random process X with a denumerable state space. On the other hand, in queueing theory a random process describing the evolution of a characteristic is a function of an m.p.p. in which points represent instants of arrival, and the mark attached to any point describes the service characteristics (as e.g. service time, priority index etc.) of the customer who arrived at this instant.

Denote for a fixed $K \in \mathcal{K}$

$$F_K^o = \{(d,n) \in F,\ n(\{0\} \times K) = 1\}.$$

We have

$$\mathcal{B}F_K^o = \{F \cap F_K^o,\ F \in \mathcal{B}F\} .$$

If $n \in N_K$ is interpreted as a collection of points of R with marks attached to them $(n \leftrightarrow \{y_i, k_i\}$, where $\{y_i, k_i\}$ are all atoms of $n)$ then $n \in F_K^o$ means that there is a point at zero with a mark in K.

Denote $\Phi\hat{} = (X\hat{}, N\hat{})$ where for $(d,n) \in F$

$$X\hat{}(1,n) = 1 , \qquad N\hat{}(1,n) = n .$$

Similarly any measurable function on F we provide with the mark $\hat{}$ (except for shift transformations τ and σ defined later on).

Let $n \in N_K$. Then, by Kallenberg (1976), Lemma 3.2

$$n = \sum_i \delta_{(y_i, k_i)}$$

and the mapping, for any $i \in Z$,

$$N_K \ni n \to (y_i, k_i) \in R \times K$$

is measurable. We can reorder the sequence $\{y_i, k_i\}$ in such a way that

$$\ldots < y_{(-1)} < y_{(0)} \leq 0 < y_{(1)} < \ldots .$$

It can be shown that the mapping

$$R:\ R^\infty \to R^\infty$$

ordering the components in ascending order is measurable, so that the mapping, for any $i \in Z$

$$N_K \ni n \to y_{(i)} \in R$$

is measurable. Denote

$$Y_i^{\wedge}(f) = y_{(i)} ,$$

$$K_i^{\wedge}(f) = k_{(i)} , \qquad i \in Z .$$

Since $(Y_i^{\wedge}, K_i^{\wedge})$ is measurable so is

$$(Y_i, K_i) = (Y_i^{\wedge}(\Phi), K_i^{\wedge}(\Phi)) , \qquad i \in Z .$$

Let $H \in \mathcal{B}K$ be fixed. From the sequence $\{y_{(i)}, k_{(i)}\}$ we cancel out all $(y_{(i)}, k_{(i)})$ such that $k_{(i)} \notin H$. Denote the resulting sequence by $\{y_{H,i}, k_{H,i}\}$. It is indexed according to the convention

$$\dots < y_{H,-1} < y_{H,0} \le 0 < y_{H,1} < \dots .$$

Denote

$$Y_{H,i}^{\wedge}(f) = y_{H,i} ,$$

$$K_{H,i}^{\wedge}(f) = k_{H,i} , \qquad i \in Z .$$

Since $(Y_{H,i}^{\wedge}, K_{H,i}^{\wedge})$ are measurable so is

$$(Y_{H,i}, K_{H,i}) = (Y_{H,i}^{\wedge}(\Phi), K_{H,i}^{\wedge}(\Phi)), \qquad i \in Z .$$

To introduce groups of shift transformations on F we must first introduce them on $\mathcal{D}(R,E)$ and N_K. The group of shift transformations $\underline{\tau}_1 = \{\tau_1^t, \ t \in R\}$ is defined as in Section 1.5 by

$$\tau_1^t d = d(t+\cdot), \ t \in R, \ d \in \mathcal{D}(R,E) .$$

From Lemma 1.1 it follows that $\underline{\tau}_1$ is a measurable group. Now let $n \in N_K$ and denote by $\tau_2^t n$ a measure from N_K which gives the mass $n(C +_1 t)$ to the set $C \in \mathcal{B}(R \times K)$ where

$$C +_1 t = \{(a+t,b), \ (a,b) \in C\} .$$

The transformation

$$R \times N_K \ni (t,n) \to \tau_2^t n \in N_K$$

is known to be measurable; thus $\underline{\tau}_2 = \{\tau_2^t,\ t \in R\}$ is a measurable group of transformations of N_K onto itself.

Definition 3.3:

On (F, BF) define the group of shift transformations $\underline{\tau} = \{\tau^t, t \in R\}$ by

$$\tau^t(d,n) = \{\tau_1^t d, \tau_2^t n\}\ , \qquad t \in R\ , \quad (d,n) \in F$$

and the group of shift transformations $\underline{\sigma}_K = \{\sigma_K^i\}$, by

$$\sigma_K^i(d,n) = (\tau_1^{\hat{Y}_{K,i}(d,n)} 1,\ \tau_2^{\hat{Y}_{K,i}(d,n)} n)\ .$$

Convention:

We do not write the subscript K when $K = K$, e.g. $Y_{K,i} = Y_i$, $\underline{\sigma}_K = \underline{\sigma}$.

Remark:

Both $\underline{\tau}$ and $\underline{\sigma}_K$ are measurable groups of transformations of F into itself. Moreover τ^t maps F one-to-one onto F, $t \in R$ and σ_K^i maps F_K^o one-to-one onto F_K^o, $i \in Z$.

Recall that according to Section 1.5 an r.e.@ m.p.p.

$$\Phi = (X,N): (A, SA, Pr) \to (F, BF)$$

is called *stationary* if for every $F \in BF$ and $t \in R$

$$Pr(\Phi \in F) = Pr(\tau^t \Phi \in F)\ .$$

In a similar way we introduce the concept that Φ is *ergodic* or *metrically-transitive*.

The following concept is basic in this set of notes.

Definition 3.4:

A distribution P_K^o on (F, BF) is said to be a *Palm distribution with respect to marks from* K if

$$P_K^o \sigma_K^{-1} = P_K^o\ .$$

An r.p.@ m.p.p. Φ_K is said to be a *Palm version with respect to marks from* K if its distribution $P_K^o = Pr(\Phi_K^o)^{-1}$ is a Palm distribution with respect to marks from K. If $K = \mathcal{K}$ then the subscript K is omitted and we say simply that Φ^o is a *Palm distribution* and Φ^o is a *Palm version.*

Proposition 3.1:

If P_K^o is a Palm distribution with respect to marks from K then

$$P_K^o(F_K^o) = 1 .$$

The proof is the same as the proof of Proposition 2.1.

Example 3.1:

Let $\{T_i^o\}$ be a stationary sequence of positive r.v.'s. Set

$$Y_j^o = \begin{cases} \sum_{i=0}^{j-1} T_i^o , & j=1,2,\ldots \\ 0 , & j=0 \\ -\sum_{i=-1}^{j} T_i^o , & j=-1,-2,\ldots . \end{cases}$$

The p.p.

$$N^o = \sum_i \delta_{Y_i^o}$$

is a Palm version.

We now present a few further examples. They serve as illustrations of the introduced concepts. Nevertheless they will also be used later for obtaining original results in various applied fields, especially in the theory of queues.

Example 3.2 (v.w.t.p.@ input m.p.p. in a G/G/s; FIFO queue):

Consider a G/G/s; FIFO queue with the generic sequence $\{T_i^o, S_i^o, K_i^o\}$. The third component of the generic sequence, i.e. $\{K_i^o\}$ indicates which class an arriving customer comes from. The input m.p.p. N^o is defined as

$$N^o = \sum_i \delta_{(Y_i^o, K_i^o)} ,$$

where

$$Y_j^o = \begin{cases} \sum_{i=0}^{j-1} T_i^o , & j=1,2,\dots \\ 0 , & j=0 \\ -\sum_{i=-1}^{j} T_i^o , & j=-1,-2,\dots . \end{cases}$$

Our aim is to study the v.w.t.p. $\tilde{V} = \{\tilde{V}_1(t),\dots,\tilde{V}_s(t),\ t \in R\}$. The first component $\{\tilde{V}_1(t)\}$ represents the work-load process in the queue called a v.w.t.p.. In the single server case we write $\tilde{V} = \{\tilde{V}(t),\ t \in R\}$. Let $\tilde{W} = \{\underline{\tilde{W}}_i,\ i \in Z\}$ be the vector a.w.t.p.. We suppose that

$$\underline{\tilde{W}}_{i+1} = \begin{cases} \underline{0} & i=-1,-2,\dots \\ R[\underline{\tilde{W}}_i + \underline{S}_i^o - \underline{T}_i^o]^+ , & i=0,1,\dots \end{cases}$$

where $\underline{S}_i^o$, $\underline{T}_i^o$ and R were defined after the formula (1.10). Denote also

$$\underline{t} = (t,\dots,t) , \qquad \underline{Y}_i^o = (Y_i^o,\dots,Y_i^o) .$$

Within the interval $Y_i \le t < Y_{i+1}^o$ the r.p. $\underline{V}^o$ is defined by

$$\underline{\tilde{V}}(t) = R[\underline{\tilde{W}}_i + \underline{S}_i^o - \underline{t} + \underline{Y}_i^o]^+ .$$

Note that

$$\underline{\tilde{V}}(Y_i^o - o) = \underline{W}_i^o , \qquad i \in Z .$$

To define a Palm representation of a v.w.t.p.@ input p.p. we need to assume that $\{T_i^o, S_i^o, K_i^o\}$ is metrically-transitive and

$$\frac{E\, S_0^o}{s\, E\, T_0^o} < 1 . \tag{3.2}$$

Moreover $T_i^o > 0$, $ET_i^o < \infty$ will be needed later. Because of (3.2) there exists a finite stationary sequence $\{\underline{M}_i^o\}$ fulfilling (1.10). We assume that $\{\underline{M}_i^o\}$ is a minimal stationary sequence in the sense of Loynes (1962). Thus for consistency, in the multivariate case $(s > 1)$ we must assume that $\underline{\tilde{W}}_0 = \underline{0}$. This is important because otherwise, to a sequence $\{\tilde{W}_i\}$ not fulfilling the condition $\underline{\tilde{W}}_0 = 0$, a stationary sequence other than $\{M_i^o\}$ may correspond. Within the interval $Y_i^o \le t < Y_{i+1}^o$ the r.p. $\underline{V}^o$ we define by

$$V^o(t) = R[\underline{M}_i^o + \underline{S}_i^o - \underline{t} + \underline{Y}_i^o]^+ .$$

Note that

$$\underline{V}^o(Y_i^o - o) = \underline{M}_i^o , \qquad i \in Z . \tag{3.3}$$

We have defined here two r.p.@ m.p.p.'s $\tilde{\Phi} = (\tilde{\underline{V}}, N^o)$ and $\Phi^o = (\underline{V}^o, N^o)$. In the next proposition we show that Φ^o is indeed a Palm version. It is called a Palm representation of the v.w.t.p. in the queue.

Proposition 3.2:
$\Phi^o = (\underline{V}^o, N^o)$ is a $\underline{\sigma}$-ergodic Palm version.

Proof: To prove the proposition we apply Proposition 1.2. Note first that $\Phi^o = \Phi^o(\{T_i^o, S_i^o, K_i^o\})$. The sequence $\{T_i^o, S_i^o, K_i^o\}$ is supposed to be metrically-transitive, so the mapping $\tau: (R \times R \times Z)^Z \to (R \times R \times Z)^Z$ defined by

$$\theta(\{t_i, s_i, k_i\}) = \{t_{i+1}, s_{i+1}, k_{i+1}\}$$

is measure preserving on

$$((R \times R \times Z)^Z, \mathcal{B}(R \times R \times Z)^Z , \text{ distribution of } \{T_i^o, S_i^o, K_i^o\}).$$

Now to complete the proof it suffices to note that

$$\sigma\Phi^o(\{T_i^o, S_i^o, K_i^o\}) = \Phi^o(\theta\{T_i^o, S_i^o, K_i^o\}) .$$

Example 3.3 (v.w.t.p.@ input m.p.p. in G/G/1; work conserving queue):
According to the definition, the virtual waiting time is identical for all work-conserving queues with the same generic sequence $\{T_i^o, S_i^o, K_i^o\}$. We define the associated m.p.p. N^o as in Example 3.2. Similarly we define the v.w.t.p. $\tilde{V}$ and a Palm representation V^o. However note that unless a queueing discipline is FIFO the sequences $\{\tilde{W}_i\}$ and $\{M_i^o\}$ cannot be interpreted as a.w.t.p.'s in the queue.

Example 3.4 (content process @ input p.p. in a dam):
The generic sequence of a dam is $\{T_i^o, S_i^o\}$. Here T_i^o, $i \in Z$ are dinstances between two consecutive instants of inpouring, while S_i^o, $i \in Z$ are amounts of the water inpoured into the dam. Let $r: R_+ \to R_+$

be a function such that $r(0) = 0$, positive and continuous on $(0,\infty)$. The function r is the release rule; namely $r(x)$ is the rate of output if the dam content is x. Notice that if $r(x) = 1$, $x > 0$ then the process of the content of the dam is simply the work-load process (v.w.t.p.) in the G/G/1; FIFO queue with the generic sequence $\{T_i^o, S_i^o\}$. We aim to study the process of the content of the dam. To do it we proceed as we did in Example 3.2 and 3.3. Between two consecutive instants of inpouring the output process is a deterministic one. It is given by a function $z(x,t)$ which shows the content after time t if at the beginning the content was x. The function $z(x,t)$ is the solution of the integral equation

$$z(x,t) = x - \int_0^t r(z(x,s))ds \,, \qquad (3.4)$$
$$z(x,0) = x \,.$$

The function z is nondecreasing of the first variable. Denote by $\tilde{W}$ a process of the content of the dam just before the instants of inpouring. The process $\tilde{W}$ fulfills the following recursive relationship

$$\tilde{W}_{i+1} = \begin{cases} 0 \,, & i=-1,-2,\ldots \\ z(\tilde{W}_i + S_i^o, T_i^o) \,, & i=0,1,\ldots \,. \end{cases}$$

The associated p.p. N^o we define by

$$N^o = \sum_i \delta_{Y_i^o} \,,$$

where $\{Y_i^o\}$ were defined in Example 3.2. Within the interval $Y_i^o \le t < Y_{i+1}^o$ the content process $\tilde{X}$ is

$$\tilde{X}(t) = z(\tilde{W}_i + S_i^o,\ t - Y_i^o) \,, \qquad t \in R \,.$$

To define a Palm version we need to assume that

(a) $\{T_i^o, S_i^o\}$ is metrically-transitive,

(b) $T_i^o > 0$, $i \in Z$.

Moreover

(c) $ET_i^o < \infty$

will be needed later on. Loynes' lemma insures the existence of a stationary sequence $\{M_i^o\}$ which fulfills

$$(3.5) \qquad M_i^o = z(M_i^o + S_i^o, T_i^o) , \qquad i \in Z .$$

We assume that $\{M_i^o\}$ is a minimal sequence. However assumed conditions (a), (b) do not insure the finitness of M_i^o, $i \in Z$. Thus we assume that $\{M_i^o\}$ is a sequence of finite r.v.'s. Within the interval $Y_i^o \le t < Y_{i+1}^o$ the r.p. X^o is defined by

$$X^o(t) = z(M_i^o + S_i^o, \ t - Y_i^o) , \qquad t \in R .$$

The proof that (X^o, N^o) is a Palm version is similar to the proof of Proposition 3.2. If there exists more than one stationary sequence fulfilling (3.5) (regarding their distributions) then, similarly as we did in Example 3.2, we must assume that $\tilde{W}_0 = 0$.

Example 3.5 (queue size process @ input m.p.p. in G/G/1; FIFO queue): Define the mapping

$$(3.6) \qquad L^{\wedge}: \mathcal{D}(R,R_+) \times N_K \to \mathcal{D}(R,R_+)$$

by

$$L^{\wedge}(v,n)(t) = \#\{i: v(y_i) + y_i > t\} .$$

Recall that $(v,n) = (v, \sum_i \delta_{(y_i,k_i)}) \in F$. A few words should be said about the measurability of $L^{\wedge}$. Denote for $(v,n) \in \mathcal{D}(R,R_+) \times N_K$

$$V^{\wedge}(v,n) = v .$$

Recalling the notation $Y_i^{\wedge}$ introduced in § 1 we rewrite $L^{\wedge}$ in the form

$$L^{\wedge}(v,n) = \sum_i 1_{(t,\infty)}(V^{\wedge}(Y_i^{\wedge}) + Y_i^{\wedge}) .$$

Bearing in mind that the mapping

$$R \times \mathcal{D}(R,R_+) \ni (t,v) \to v(t) \in R$$

is measurable, we have for any $i \in Z$, that $V^{\wedge}(Y_i^{\wedge})$ is an r.v.. So

$$\{1_{(t,\infty)}(V^{\wedge}(Y_i^{\wedge}) + Y_i^{\wedge}) , \ t \in R\}$$

is an r.p. and hence $L^\wedge$ is also. Consider a queue with the a.w.t.p. $\tilde{V}$ associated with the input m.p.p. N^o defined in Example 3.2. Then the queue size process is

$$\tilde{L}(t) = L^\wedge(\tilde{V},N^o)(t) , \qquad t \in R .$$

A Palm representation is (L^o,N^o) where the r.p. L^o is defined by

$$L^o(t) = L^\wedge(V^o,N^o)(t) , \qquad t \in R .$$

Here (V^o,N^o) is a Palm representation of the v.w.t.p.@ input m.p.p..

Example 3.6 (queue size process @ input m.p.p. in a queue):

The definition of the queue size process given in the previous example may not be valid for other than FIFO disciplines. Here we give a definition of the queue size process which works for all disciplines. Let $\{T_i^o,S_i^o,K_i^o\}$ be a generic sequence where:

T_i^o denotes the inter-arrival time between the i-th and (i+1)-st customer,

S_i^o denotes the service time of the i-th customer,

K_i^o denotes the class which the i-th customer comes from.

Let $\{\tilde{W}_i\}$ be the a.w.t.p. in the queue. As usual we complete the single-ended sequence to the double-ended one putting

$$\tilde{W}_i = 0 , \qquad i=-1,-2,\ldots .$$

The sojourn time of the i-th customer is

$$\tilde{V}_i = \begin{cases} 0 , & i=-1,-2,\ldots \\ \tilde{W}_i + S_i^o , & i=0,1,\ldots . \end{cases}$$

Set

$$Y_j^o = \begin{cases} \sum_{i=1}^{j} T_i^o , & j=1,2,\ldots \\ 0 , & j=0 \\ -\sum_{i=-1}^{j} T_i^o , & j=-1,-2,\ldots . \end{cases}$$

The queue size process is

$$\tilde{L}(t) = \#\{i,\ Y_i^o \le t < Y_i^o + \tilde{V}_i\}$$

and the associated m.p.p. is

$$N^o = \sum_i \delta_{(Y_i^o, K_i^o)} .$$

To define a Palm version of the queue size we must assume that there exists an a.w.t.p. $\{M_i^o\}$ such that

$$\{T_i^o, S_i^o, K_i^o, M_i^o\}$$

is metrically-transitive. Moreover we suppose that $\{T_i^o, S_i^o, S_i^o, \tilde{W}_i\}$ is stable and the stationary distribution of $\{T_i^o, S_i^o, K_i^o, \tilde{W}_i\}$ is identical with the distribution of $\{T_i^o, S_i^o, K_i^o, M_i^o\}$. We denote the stationary sojourn time process by $\{V_j^o\} = \{M_j^o + S_j^o\}$. Define the r.p. L^o by

$$L^o(t) = \#\{i,\ Y_i^o \le t < Y_i^o + V_i^o\} , \qquad t \in R .$$

It can be proved that (L^o, N^o) is a $\underline{\sigma}$-ergodic Palm version. It is a *Palm representation of the queue size process.*

Example 3.7 (performance process):

The following example is of reliability interest. At any time t a considered system may work or be repaired. The consecutive working times are $T_{1,i}^o$, $i \in Z$ and repairing times are $T_{2,i}^o$, $i \in Z$. Thus the renewal instants are

$$Y_{1,i}^o = \begin{cases} \sum_{i=0}^{j-1} (T_{1,j}^o + T_{2,j}^o) , & j=1,2,\dots \\ 0 , & j=0 \\ -\sum_{i=-1}^{j} (T_{1,j}^o + T_{2,j}^o) , & j=-1,-2,\dots . \end{cases}$$

and the failure instants are

$$Y_{2,i}^o = Y_{1,i}^o + T_{1,i}^o , \qquad i \in Z .$$

The performance process X^o indicates a state of the system at time t . We have

$$X^{\circ}(t) = \begin{cases} 1 \ , & Y^{\circ}_{1,j} \le t < Y^{\circ}_{2,j} \ , \quad \text{for some} \quad j \in Z \\ 0 \ , & Y^{\circ}_{2,j} \le t < Y^{\circ}_{1,j} \ , \quad \text{for some} \quad j \in Z \ . \end{cases}$$

Set

$$N^{\circ} = \sum_j \delta_{(Y^{\circ}_{1,j},1)} + \sum_j \delta_{(Y^{\circ}_{2,j},2)} = \sum_j \delta_{(Y^{\circ}_j,K^{\circ}_j)} \ .$$

Clearly K°_j describes whether Y°_j is a failure or a renewal instant. From now on we assume that $\{T^{\circ}_{1,i},T^{\circ}_{2,i}\}$ is metrically-transitive and that

$$E(T^{\circ}_{1,0} + T^{\circ}_{2,0}) < \infty \ .$$

It turns out that (X°,N°) is a Palm vesion with respect to mark 1 but it is not a Palm version.

§ 2. Construction of stationary r.p.@ m.p.p.'s.

As in the discrete case having an r.p.@ m.p.p.

$$\Phi^{\circ} = (X^{\circ},N^{\circ}): (A,SA,Pr) \to (F,BF)$$

we ask for

(3.7) $$\lim_{t\to\infty} \frac{1}{t} \int_0^t Pr(\tau^s\Phi^{\circ} \in F)ds = P(F) \ , \qquad F \in BF \ .$$

If this limit exists and is an honest distribution on F then we call it the *stationary distribution of the* r.p.@ m.p.p. Φ°. If Φ° is a Palm version we can formally define P as a function of the distribution of Φ° and then verify that (3.7) indeed holds. In this section we investigate properties of the stationary distribution corresponding to a Palm distribution. Ergodic theorems, among others a theorem asserting (3.7) holds, are given in the next section.

In what follows Φ°_K denotes a Palm version with respect to marks from K and P°_K denotes the distribution of Φ°_K. Denote

$$\lambda_K^{-1} = \int_F \hat{Y}_{K,1}(f) P_K^o(df) \ .$$

It is a general assumption in these notes that the set K is such that

$$0 < \lambda_K < \infty \ .$$

For any $F \in \mathcal{B}F$ define

(3.8) $$P(F) = \lambda_K \int_F (\int_0^{\hat{Y}_{K,1}(f)} 1_F(\tau^t f) dt) P_K^o(df) \ .$$

Proposition 3.3:
P is a stationary distribution on F.

Proof: Clearly P is a distribution on F. Let $F \in \mathcal{B}F$. Then for any $t \in R$

(3.9) $$P(\tau^{-t}F) = \lambda_K \int_F (\int_0^{\hat{Y}_{K,1}(f)} 1_F(\tau^{s+t} f) ds) P_K^o(df)$$

$$= \lambda_K \int_F (\int_t^{t+\hat{Y}_{K,1}(f)} 1_F(\tau^s f) ds) P_K^o(df)$$

$$= \lambda_K \int_F (\int_0^{\hat{Y}_{K,1}(f)} 1_F(\tau^s f) ds) P_K^o(df)$$

$$+ \lambda_K \int_F (\int_{\hat{Y}_{K,1}(f)}^{\hat{Y}_{K,1}(f)+t} 1_F(\tau^s f) ds - \int_0^t 1_F(\tau^s f) ds) P_K^o(df) \ .$$

Since $\{\tau^s, s \in R\}$ is a measurable group we have that

$$I(f) = \int_0^t 1_F(\tau^s f) ds \ , \quad \text{any} \quad t \geq 0$$

is an r.v.. Clearly

$$I(\sigma_K f) = \int_{\hat{Y}_{K,1}(f)}^{\hat{Y}_{K,1}(f)+t} 1_F(\tau^s f) ds \ , \qquad f \in F_K^o \ .$$

Since P_K^o is a Palm distribution we have that I is identically distribution as $I\sigma_K$. Thus

$$\int_F \int_0^t 1_F(\tau^s f)ds\, P_K^o(df) = \int_F \int_{Y_{K,1}^{\wedge}(f)}^{Y_{K,1}^{\wedge}(f)+t} 1_F(\tau^s f)ds\, P_K^o(df) ,$$

and hence by (3.9) we obtain $P(\tau^{-t}F) = P(F)$.

Proposition 3.4:

For any P-integrable function φ: $F \to R$

$$(3.10) \qquad \int_F \varphi(f)P(df) = \lambda_K \int_F \Big(\int_{Y_{K,i}^{\wedge}(f)}^{Y_{K,i+1}^{\wedge}(f)} \varphi(\tau^s f)ds\Big)P_k^o(df) , \qquad i \in Z .$$

Proof: For $i=0$, the formula (3.10) is obtained from (3.8) by the standard approximation technique of any function φ by simple functions. The general case follows from the case just proved $i=0$ and from that

$$I_i(f) = \int_0^{T_{K,i}^{\wedge}(f)} \varphi(\tau^{Y_{K,i}^{\wedge}(f)+s} f)ds ,$$

(where $T_{K,i}^{\wedge} = Y_{K,i+1}^{\wedge} - Y_{K,i}^{\wedge}$, $i \in Z$) are identically distributed on (F_K^o, BF_K^o, P_K^o).

Let $n \in N_K$ and $K \in BK$. Define n_K^*: $N_K \to N(R)$ by

$$n_K^*: n \to n(\cdot \times K) .$$

The mapping n_K^* is measurable. To demonstrate this we recall that $BN(R)$ is generated by sets of a form $\underline{M} = \{n_K^*: n_K(B) = k\}$, $k \in N_0$, $B \in BR$. Now

$$\{n, n_K^*(n) \in \underline{M}\} = \{n, n(B \times K) = k\} \in BN_K .$$

The identity given in the next proposition is basic in these notes. It is a generalization of Mecke's onto the case of r.p.@ m.p.p.'s.

Proposition 3.5 (Mecke):

For any measurable function φ: $F \times R \to R_+$

$$(3.11) \qquad \int_F \int_R \varphi(f,t)n_K^*(dt)P(df) = \lambda_K \int_F \int_R \varphi(\tau^t f, -t)dt\, P_K^o(df) .$$

Proof: We have

$$\int_F \int_R \varphi(f,t)n_K^*(dt)P(df) = \int_F \sum_i \varphi(f,Y_{K,i}^{\wedge}(f))P(df) .$$

By (3.10), bearing in mind that P_K^o is a Palm distribution,

$$\int_F \sum_i \varphi(f,Y_{K,i}^{\wedge}(f))P(df)$$

$$= \sum_i \lambda_K \int_F \int_{Y_{K,i}^{\wedge}(f)}^{Y_{K,i+1}^{\wedge}(f)} \varphi(\tau^s f,Y_{K,i}^{\wedge}(\tau^s f))ds\ P_K^o(df)$$

$$= \sum_i \lambda_K \int_F \int_{Y_{K,0}^{\wedge}(\sigma_K^i f)}^{Y_{K,1}^{\wedge}(\sigma_K^i f)} \varphi(\tau^s f,Y_{K,0}^{\wedge}(\tau^s \sigma_K^i f))ds\ P_K^o(df)$$

$$= \sum_i \lambda_K \int_F \int_{Y_{K,0}^{\wedge}(f)}^{Y_{K,1}^{\wedge}(f)} \varphi(\tau^s \sigma_K^{-i} f,Y_{K,0}^{\wedge}(\tau^s f)ds\ P_K^o(df)$$

$$= \sum_i \lambda_K \int_F \int_{Y_{K,0}^{\wedge}(f)}^{Y_{K,1}^{\wedge}(f)} \varphi(\tau^s \sigma_K^{-i} f, -s)ds\ P_K^o(df)$$

$$= \sum_i \lambda_K \int_F \int_{Y_{K,i}^{\wedge}(f)}^{Y_{K,i+1}^{\wedge}(f)} \varphi(\tau^s f, -s)ds\ P_K^o(df)$$

$$= \int_F \int_R \varphi(\tau^s f, -s)ds\ P_K^o(df) .$$

The following definition will be justified by Theorem 3.1 and 3.2.

<u>Definition</u> 3.4:

(i). If Φ_K^o is an r.p.@ m.p.p. with the distribution P_K^o then P defined by (3.8) is called the *stationary distribution* of Φ_K^o.

(ii). Any r.p.@ m.p.p. $\Phi = (L,N)$ with the distribution P is called a *stationary* r.p.@ m.p.p. *corresponding to* Φ_K^o.

(iii). Any r.p.@ m.p.p. Φ_K^o with the distribution P_K^o is called a *Palm version corresponding to the stationary* r.p.@ m.p.p. Φ .

<u>Example</u> 3.8:

Consider sequences $\{T_{1,i}^o\},\ldots,\{T_{l,i}^o\}$ of positive r.v.'s such that $ET_{2,i}^o < \infty,\ldots,ET_{l,i}^o < \infty$. Eeach sequence is metrically-transitive and they are independent of each other. Set for $k=1,2,\ldots,l$

$$Y^o_{k,j} = \begin{cases} \sum_{i=1}^{j-1} T^o_{k,i} , & j=1,2,\dots \\ 0 , & j=0 \\ -\sum_{i=-1}^{j} T^o_{k,i} , & j=-1,-2,\dots . \end{cases}$$

Let N_k be a stationary m.p.p. corresponding to the Palm version $N^o_k = \sum_j \delta_{(Y^o_{kj}, k)}$, $k=1,\dots,l$. Notice that the space of marks of the m.p.p. N^o_k is $K = \{k\}$. We may assume that $N^o_1, N_2, \dots, N_k$ are independent. Then the m.p.p.

$$N^o_{\{1\}} = N^o_1 + N_2 + \dots + N_l$$

is a Palm version with respect to marks from $\{1\}$. To demonstrate this fact we find out that N^o_1 is distributed identically to $\tau^{Y^o_{1,1}} N^o_1$ and N_k is distributed identically to $\tau^{Y^o_{1,1}} N_k$, $k=2,\dots,l$. Recall that $Y^o_{1,1}$, $N_2,\dots,N_k$ are independent. Hence $N^o_{\{1\}}$ is distributed identically to $\sigma N^o_{\{1\}} = \tau^{Y_{1,1}} N^o_{\{1\}}$.

Example 3.9:

Consider $\{T^o_j\}$ a stationary independent sequence of positive r.v.'s. Assume that $ET^o_0 < \infty$. Set

$$Y^o_j = \begin{cases} \sum_{i=1}^{j-1} T^o_i , & j=1,2,\dots \\ 0 , & j=0 \\ -\sum_{i=-1}^{j-1} T^o_i , & j=-1,-2,\dots . \end{cases}$$

Let N be a stationary sequence corresponding to the p.p. $N^o = \sum_i \delta_{Y^o_i}$. Denote by $T_1, T_2, \dots$ the consecutive inter-point distances on $(0,\infty)$ (T_1 is the distance to the nearest point to the right of zero). Then $\{T_2, T_3, \dots\}$ is also a stationary independent sequence distributed identically to $\{T^o_1, T^o_2, \dots\}$. This fact results because using (3.8) for $k=1,2,\dots$, $B_1,\dots,B_k \in BR_+$ we have

$$Pr(T_2 \in B_1,\dots,T_{k+1} \in B_k) = Pr(T^o_1 \in B_1,\dots,T^o_k \in B_k) .$$

In the next theorem $\Phi = (X,N)$ is a stationary r.p.@ m.p.p. corresponding to Φ_K^o.

Theorem 3.1 (Ryll-Nardzewski):

(i) $\lambda_K = EN([0,1] \times K)$,

(ii) For any stationary distribution P there exists a unique Palm distribution P_K^o with respect to marks from K such that (3.8) holds. Moreover for any Borel function $g: R \to R_+$ such that $\int_R g(t)dt = 1$ we have

$$(3.12) \quad P_K^o(F) = \left(\int_F \int_R g(t)n_K^*(dt)P(df)\right)\left(\int_F \int_R 1_F(\tau^{-t}f)g(t)n_K^*(dt)P(df)\right) .$$

Proof: (i). Set $\varphi(f) = n([0,1] \times K)$ in (3.10).
(ii). To get (3.12) set in (3.10)

$$\varphi(f,t) = 1_F(\tau^{-t}f)g(t) , \qquad t \in R , \quad F \in BF .$$

The uniqueness results from the following. If there exist two Palm distributions P_K^o and P_K^* (with respect to marks from K) corresponding to P, then for any P-integrable function φ, we have by (3.8)

$$\frac{\int_F \int_0^{Y_{K,1}^{\wedge}(f)} \varphi(\tau^s f)ds\, P_K^o(df)}{\int_F Y_{K,1}^{\wedge}(f)P_K^o(df)} = \frac{\int_F \int_0^{Y_{K,1}^{\wedge}(f)} \varphi(\tau^s f)ds\, P_K^*(df)}{\int_F Y_{K,1}^{\wedge}(f)P_K^*(df)} .$$

Let $\Psi: (F,BF) \to (R_+,BR_+)$. Set

$$\varphi(f) = \frac{\psi(\sigma_K^{-1} f)}{Y_{k,1}(f) - Y_{K,0}(f)}$$

Then for any ψ

$$\frac{\int_F \psi(f)P_K^o(df)}{\int_F Y_{K,1}^{\wedge}(f)P_K^o(df)} = \frac{\int_F \psi(f)P_K^*(df)}{\int_F Y_{K,1}^{\wedge}(f)P_K^*(df)} ,$$

which yields

$$\int_F \psi(f)P_K^o(df) = \int_F \psi(f)P_K^*(df) .$$

This shows the uniqueness and the proof is completed.

Remarks:

From Theorem 3.1 (i) it follows that λ_K is the *intensity of points with marks in* K. One may also ask for an interpretation of P_K^o. Let $\varphi: F \to R$ be a bounded measurable function, such that for any $f \in F$ the function $\varphi(\tau^t f)$ is continuous in t. Then

$$(3.13) \qquad \lim_{t \downarrow 0} \frac{1}{t} \int_{\{n([0,t]\times K)=1\}} \varphi(f)P(df) = \lambda_K \int_F \varphi(f)P_K^o(df)$$

or in another form if $\Phi = (X,N)$ is a stationary r.p.@ m.p.p. corresponding to Φ_K^o a Palm version with respect to marks from K then

$$(3.14) \qquad \lim_{t\downarrow 0} \frac{1}{t} E(\varphi(\Phi)|N([0,t]\times K) = 1) = E\varphi(\Phi_K^o) \qquad (= \int_F \varphi(f)P_K^o(df)) .$$

The proofs of (3.13) and (3.14) are identical to the proof of Theorem 6 of Ryll-Nardzewski (1961). From (3.14) we can say that P_K^o is a "conditional distribution of Φ given the condition that there is a point at zero with a mark in K".

Let $\Phi^o = (X^o, N^o)$ be a Palm version. Recall that $N^o = \sum_i \delta_{(Y_i^o, K_i^o)}$.

Proposition 3.6:

For any $H \in \mathcal{B}K$

$$(3.15) \qquad \Pr(K_0^o \in H) = \frac{\lambda_H}{\lambda} .$$

Proof: Let $F_H = \{f: k_0 \in H\}$, where $F \ni f = (1,n) = (1, \sum_i \delta_{(y_i,k_i)})$. Then

$$\Pr(K_0^o \in H) = P^o(F_H)$$

and by (3.12), setting $g(t) = 1_{[0,1]}(t)$

$$P^o(F_H) = \lambda^{-1} \int_F \int_0^1 1_F(\tau^{-t} f)n^*(dt)P(df) = \lambda^{-1} \int_F \int_0^1 n_H^*(dt)P(df) = \frac{\lambda_H}{\lambda} .$$

Proposition 3.7:

For any $H \in \mathcal{B}K$ such that $\lambda_H > 0$

$$(3.16) \qquad \Pr(\Phi^o \in \cdot \,|\, K_0^o \in H) = P_H^o(\cdot) .$$

Proof: Let $F \in \mathcal{BK}$ and $F_H = \{f: k_0 \in H\}$. Then

$$Pr(\Phi^o \in F | K_0^o \in H) = P^o(F|F_H)$$

and

$$P^o(F|F_H) = \frac{P^o(F \cap F_H)}{P^o(F_H)} .$$

Now by (3.12) and Proposition 3.6

$$\frac{P^o(F \cap F_H)}{P^o(F_H)} = \lambda_H^{-1} \int_F \int_0^1 1_{F \cap F_H}(\tau^{-t} f) n^*(dt) P(df)$$

$$= \lambda_H^{-1} \int_F \int_0^1 1_F(\tau^{-t} f) n_H^*(dt) P(df) = P_H^o(F) .$$

Example 3.10 (Little formula):

The assumptions and notations of Example 3.6 are in force. A stationary r.p.@ m.p.p. corresponding to (L^o, N^o) we denote by (L,N). We are going to prove the Little formula

$$EL(0) = \lambda E V_0^o , \tag{3.17}$$

where $\lambda^{-1} = ET_0^o$. By Proposition 3.6 and 3.7 we get

$$EL(0) = \sum_{i \in K} \lambda_i E(V_0^o | K_0^o = i) . \tag{3.18}$$

Here $E(V_0^o | K_0^o = i)$ can be read as the stationary sojourn time of a customer from the i-th class. To prove (3.17) consider the m.p.p.

$$\bar{N}^o = \sum_i \delta_{(Y_i^o, [V_i^o, K_i^o])} .$$

It can be easily proved that $\bar{N}^o$ is a Palm version. The stationary m.p.p. corresponding to $\bar{N}^o$ we denote by $\bar{N}$. For $n = \sum_i \delta_{(y_i, [v_i, k_i])} \in$ $\in N_{R_+ \times Z}$, $t \in R$ set

$$L^\wedge(n,t) = \#\{i,\ y_i \le t < y_i + v_i\} , \qquad t \in R .$$

It can be proved that $L^{\wedge}: N_{R_+ \times Z} \to \mathcal{D}(R, R_+)$ is measurable. Note that

$$L^o(t) = L^{\wedge}(\overline{N}^o, t)$$

and that L and $L^{\wedge}(\overline{N})$ are identically distributed. For $n \in N_{R_+ \times Z}$, $t \in R$ set

$$(3.19) \qquad \varphi(n,t) = \begin{cases} 1, & \text{for some } i,\ y_i \le 0 < y_i + v_i,\ t = y_i \\ 0, & \text{otherwise}. \end{cases}$$

We have for $n \in N_{R_+ \times Z}$

$$\int_R \varphi(n,t) n(dt \times R_+ \times Z) = L^{\wedge}(n,0)$$

and for n having a point at zero

$$\int_R \varphi(\tau_2^t n, -t) dt = v_0 .$$

Applying Mecke's identity (3.11) with φ defined in (3.19) we obtain

$$EL^{\wedge}(\overline{N},0) = \lambda E V_0^o .$$

This completes the proof because

$$EL^{\wedge}(\overline{N},0) = EL(0) .$$

§ 3. Ergodic theorems.

In this section we show that P defined by (3.8) is indeed the sought for stationary distribution of Φ^o. It will be given in Corollary 3.2 (i). In Theorem 3.3 we also positively solve a more general problem i.e. whether the $\underline{\sigma}$-stability of an r.p.@ m.p.p. $\tilde{\Phi}$ ($\tilde{\Phi}$ need not be a Palm version) yields the stability of $\tilde{\Phi}$. By $\underline{\sigma}$-stability we mean that

$$\lim_{j \to \infty} \sum_{i=0}^{j-1} \Pr(\sigma^i \tilde{\Phi} \in \cdot) = P^o(\cdot) ,$$

where P^o is a Palm distribution on F. Recall that $K \in \mathcal{B}K$ is such that $0 < \lambda_K < \infty$.

Theorem 3.2:

(i). Let Φ_K^o be a σ-ergodic Palm version with respect to marks from K defined on $(A, SA, \Pr)$. If $\varphi\colon F \to R$ is a P-integrable function then

$$(3.20) \qquad \lim_{t\to\infty} \frac{1}{t} \int_0^t \varphi(\tau^s \Phi_K^o)ds = E\varphi(\Phi), \qquad \text{a.e.,}$$

where Φ is a stationary r.p.@ m.p.p. corresponding to Φ^o. Moreover if G is a Polish space and $\psi\colon F \to G$ is a measurable mapping then for $\mu = \Pr\psi(\Phi_K^o)^{-1}$ - a.e. points $x \in G$

$$(3.21) \qquad \lim_{t\to\infty} \frac{1}{t} \int_0^t \varphi(\tau^s \Phi_K^o)ds = E\varphi(\Phi), \qquad \Pr(\cdot\,|\,\psi(\Phi_K^o) = x) \text{ - a.e..}$$

(ii). Let Φ be a metrically-transitive r.p.@ m.p.p. defined on a probability space $(A, SA, \Pr)$. If $\varphi\colon F \to R$ is a P_K^o-integrable function then

$$\lim_{j\to\infty} \frac{1}{j} \sum_{i=0}^{j-1} \varphi(\sigma_K^i \Phi) = E\varphi(\Phi_K^o), \qquad \text{a.e.,}$$

where Φ_K^o is a Palm version corresponding to Φ. Moreover if G is a Polish space and $\psi\colon F \to G$ is a measurable mapping then for $\mu = \Pr\psi(\Phi)^{-1}$ - a.e. points $x \in G$

$$\lim_{j\to\infty} \frac{1}{j} \sum_{i=0}^{j-1} \varphi(\sigma_K^i \Phi) = E\varphi(\Phi_K^o), \qquad \Pr(\cdot\,|\,\psi(\Phi) = x) \text{ - a.e..}$$

The statement remains unchanged if we set Φ_L^o in place of Φ, where $L \in \mathcal{B}K$ is such that $0 < \lambda_L < \infty$.

The proof of Theorem 3.2 we precede by two lemmas.

Lemma 3.1:

Let F be an invariant set in $(F, \mathcal{B}F)$. Then $P(F) = 1$ if and only if $P_K^o(F) = 1$.

Proof: We use (3.8) and (3.12) and proceed as in the proof of Lemma 2.1.

Denote by Φ a stationary r.p.@ m.p.p. corresponding to Φ_K^o.

Lemma 3.2:

Let $K \in \mathcal{BK}$ be such that $0 < \lambda_K < \infty$. Φ is ergodic if and only if Φ_K^o is $\underline{\sigma}_K$-ergodic.

Proof: The proof is identical to the proof of Lemma 2.2.

Proof of Theorem 3.2: (i). We use the continuous version of the ergodic theorem and proceed identically as in the proof of Theorem 2.2 (i). We may use the ergodic theorem because $\underline{\tau}$ is a measurable group. (ii). The proof is identical to the proof of Theorem 2.2 (ii).

Corollary 3.2:

Let $K \in \mathcal{BK}$ be such that $0 < \lambda_K < \infty$.

(i). For any $F \in \mathcal{BF}$

$$(3.22) \qquad \lim_{t\to\infty} \frac{1}{t} \int_0^t \Pr(\tau^s \Phi_K^o \in F)ds = P(F) .$$

Moreover for $\mu = \Pr\psi(\Phi_K^o)^{-1}$ - a.e. points $x \in G$

$$(3.23) \qquad \lim_{t\to\infty} \frac{1}{t} \int_0^t \Pr(\tau^s \Phi_K^o \in F \mid \psi(\Phi_K^o) = x)ds = P(F) .$$

(ii). For any $F \in \mathcal{BF}$

$$\lim_{j\to\infty} \frac{1}{j} \sum_{i=0}^{j-1} \Pr(\sigma_K^i \Phi \in F) = P_K^o(F) .$$

Moreover for $\mu = \Pr\psi(\Phi)^{-1}$ - a.e. points $x \in G$

$$\lim_{j\to\infty} \frac{1}{j} \sum_{i=0}^{j-1} \Pr(\sigma_K^i \Phi \in F \mid \psi(\Phi) = x)ds = P_K^o(F) .$$

The statement remains unchanged if we set Φ_L^o in place of Φ, where $L \in \mathcal{BK}$ is such that $0 < \lambda_L < \infty$.

The proof is omitted. From (3.22) we get

$$\lim_{t\to\infty} \frac{1}{t} \int_0^t \Pr(X_K^o(s) \in B)ds = \Pr(X(0) \in B) , \quad B \in \mathcal{BE}$$

or

$$\lim_{t\to\infty} \frac{1}{t} \int_0^t \Pr(X_K^o(s) \in B_1, X_K^o(s+h) \in B_2)ds = \Pr(X(0) \in B_1, X(h) \in B_2) ,$$

$$B_1, B_2 \in \mathcal{BE} .$$

We have also for any $\varphi: F \to R_+$

$$\lim_{t\to\infty} \frac{1}{t} \int_0^t \varphi(X_K^o(s))ds = E\varphi(X(0)) \quad , \qquad \text{a.e..} \tag{3.24}$$

Remark:

Consider an r.p.@ m.p.p. Φ_K^o. We assume that Φ_K^o is a $\underline{\sigma}_K$-ergodic. Palm version with respect to marks from K, $0 < \lambda_K < \infty$. From Corollary 3.2 (i) it follows that $\Phi_K^o = (L_K^o, N_K^o)$ and $\Phi = (L,N)$ are equivalent. Denote by $\{T_{K,i}^o, i \in Z\}$, $\{T_{K,i}, i \in Z\}$ sequences of consecutive inter--point distances between points with a mark at K in N_K^o and N_K respectively. We adopt the convention that the distance between the first positive and last non-positive point is indexed by 0. Then the sequences $\{T_{K,i}^o\}$ and $\{T_{K,i}\}$ are equivalent as r.e.'s on R_+^Z with the shift transformation θ given by $\theta\{x_i\} = \{x_{i+1}\}$. This assertion follows from Corollary 3.2 (ii).

The question emerges whether the $\underline{\sigma}_K$-stability of an r.p.@ m.p.p. $\tilde{\Phi}$ yields $\underline{\tau}$-stability of $\tilde{\Phi}$. If $\tilde{\Phi}$ is a $\underline{\sigma}_K$-ergodic Palm version with respect to a set K then a positive answer is given by Corollary 3.2 (i). To give a positive answer in a general case we prove two lemmata; the first one is complementary to Proposition 1.2. In the sequel $T = Z$ or R.

Lemma 3.3:

Let (G_i, SG_i), $i=1,2$ be measurable spaces and $\underline{\theta}_i = \{\theta_i^t, t \in T\}$ measurable groups of transformations of G_i, $i=1,2$ into itself respectively. Let

$$\varphi: G_1 \to G_2$$

be a measurable mapping such that

$$\varphi\theta_1^t = \theta_2^t\varphi \,, \qquad t \in T \,.$$

If an r.e.

$$X: (A, SA, Pr) \to (G_1, SG_1)$$

is stable with a stationary (ergodic) distribution P_1 then the r.e. $Y = \varphi(X)$ is stable with the stationary (ergodic) distribution $P_2 = P_1\varphi^{-1}$.

Proof: We demonstrate the proof in the case $T = R$. The remaining case $T = Z$ may be analysed similarly. For any $E_2 \in SG_2$

$$P_2(E_2) = \lim_{t\to\infty} \frac{1}{t} \int_0^t Pr(\theta_2^s Y \in E_2)ds = \lim_{t\to\infty} \frac{1}{t} \int_0^t Pr(\theta_2^s \varphi(X) \in E_2)ds$$

$$= \lim_{t\to\infty} \frac{1}{t} \int_0^t Pr(\theta_1^s X \in \varphi^{-1} E_2)ds = P_1\varphi^{-1}(E_2) ,$$

which proves the stability of Y. Now suppose that P_1 is ergodic and that $E_2 \in SG_2$ is a $\underline{\theta}_2$-invariant set. Then

$$\theta_1^{-t}\varphi^{-1}(E) = (\varphi\theta_1^t)^{-1}(E) = (\theta_2^{-t}\varphi)^{-1}(E) = \varphi^{-1}(\theta_2^{-t}E) = \varphi^{-1}(E) ,$$

which demonstrate that $\varphi^{-1}(E)$ is a $\underline{\theta}_1$-invariant. Thus

$$P_2(E) = P_1\varphi^{-1}(E) = 0 \quad \text{or} \quad 1$$

which proves that P_2 is ergodic.

Lemma 3.4:

Let (G,SG) be a measurable space and $\underline{\theta} = \{\theta^t, t \in T\}$ a measurable group of transformations of G into itself. If

$$\tilde{\Phi}: (A,SA,Pr) \to (G,SG)$$

is a stable r.e. with a stationary ergodic distribution P then for any measurable mapping $\varphi: F \to R_+$

$$\lim_{t\to\infty} \frac{1}{t} \int_0^t \varphi(\theta^s\tilde{\Phi})ds = \int_F \varphi(f)P(df) , \qquad \text{a.e. (if } T = R)$$

or

$$\lim_{j\to\infty} \frac{1}{j} \sum_{i=0}^{j-1} \varphi(\theta^i\tilde{\Phi}) = \int_F \varphi(f)P(df) , \qquad \text{a.e. (if } T = Z) .$$

Proof: We demonstrate the proof in the case $T = R$. Let Φ be an r.e. with the distribution P. From the ergodic theorem

$$\lim_{t\to\infty} \frac{1}{t} \int_0^t \varphi(\theta^s\Phi)ds = \int_G \varphi(f)P(df) , \qquad \text{a.e..}$$

Denote by $\tilde{P}$ the distribution of $\tilde{\Phi}$. If $E \in SG$ is an invariant set then

$$P(E) = \lim_{t\to\infty} \frac{1}{t} \int_0^t \tilde{P}(\theta^{-s}E)ds = \tilde{P}(E) .$$

Thus for any invariant set E, $P(E)=1$ if and only if $\tilde{P}(E)=1$. Hence

$$\lim_{t\to\infty} \frac{1}{t} \int_0^t \varphi(\theta^s\tilde{\Phi})ds = \int_G \varphi(f)P(df) , \quad \text{a.e.}$$

because the set

$$\{g \in G, \lim_{t\to\infty} \frac{1}{t} \int_0^t \varphi(\theta^s g)ds = \int_G \varphi(f)P(df)\}$$

is invariant.

In the next theorem we consider an r.p.@ m.p.p. $\tilde{\Phi} = (\tilde{X},\tilde{N})$. Denote by $\tilde{Y}_{K,i}$, $i \in Z$ the consecutive coordinates of points of $\tilde{N}$.

Theorem 3.3:

Suppose that for any $F \in BF$ there exists a $\underline{\sigma}_K$-ergodic Palm distribution such that

$$\lim_{j\to\infty} \frac{1}{j} \sum_{i=0}^{j-1} Pr(\sigma_K^i\Phi \in F) = P_K^o(F)$$

and $0 < \lambda_K < \infty$. Then for any $F \in BF$

$$P(F) = \lim_{t\to\infty} \frac{1}{t} \int_0^t Pr(\tau^s\tilde{\Phi} \in F)ds = \lambda_K \int_F (\int_0^{\hat{Y}_{K,1}(f)} 1_F(\tau^s f)ds)P_K^o(df) .$$

Proof: Define the mapping $\varphi: F \to R^Z$ by

$$\varphi(f) = \{ \int_{\hat{Y}_{K,i}(f)}^{\hat{Y}_{K,i+1}(f)} 1_F(\tau^s f)ds , \; i \in Z\} .$$

Since the mapping $R \times F \ni (s,f) \to \tau^s f \in F$ is measurable (it was shown in § 1) we have that φ is measurable. Moreover

$$\varphi\sigma_K = \theta\varphi ,$$

where $\theta: R^Z \to R^Z$ is defined by $\theta\{x_i\} = \{x_{i+1}\}$. Thus by Lemma 3.3 it follows that the sequence $\varphi(\tilde{\Phi})$ is stable with an ergodic stationary distribution. Hence, by Lemma 3.4

$$(3.25) \qquad \lim_{j\to\infty} \frac{1}{j} \sum_{i=0}^{j-1} \int_{\tilde{Y}_{K,i}}^{\tilde{Y}_{K,i+1}} 1_F(\tau^s\tilde{\Phi})ds = \int_F \left(\int_0^{Y^\wedge_{K,1}(f)} 1_F(\tau^s f)ds\right) P^o_K(df) , \quad \text{a.e..}$$

Set

$$\tilde{\nu}(t) = \max\{j: \tilde{Y}_{K,j} \le t\} , \qquad t \in R .$$

Since, by Lemma 3.4,

$$\lim_{j\to\infty} \frac{\tilde{Y}_{K,j}}{j} = EY^o_{K,1} = \lambda_K^{-1}, \qquad \text{a.e.}$$

we have

$$(3.26) \qquad \lim_{t\to\infty} \frac{\tilde{\nu}(t)}{t} = \lambda_K , \qquad \text{a.e.}$$

and

$$(3.27) \qquad \lim_{t\to\infty} \frac{t - \tilde{Y}_{K,\tilde{\nu}(t)}}{t} = 0 , \qquad \text{a.e..}$$

Below, the second equality follows from (3.27), the third from (3.25) and (3.26):

$$\lim_{t\to\infty} \frac{1}{t} \int_0^t 1_F(\tau^s\tilde{\Phi})ds =$$

$$= \lim_{t\to\infty} \frac{1}{t} \sum_{i=0}^{\tilde{\nu}(t)-1} \int_{\tilde{Y}_{K,i}}^{\tilde{Y}_{K,i+1}} 1_F(\tau^s\tilde{\Phi})ds + \lim_{t\to\infty} \frac{1}{t} \int_{\tilde{Y}_{K,\tilde{\nu}(t)}}^{t} 1_F(\tau^s\tilde{\Phi})ds$$

$$= \lim_{t\to\infty} \frac{\tilde{\nu}(t)}{t} \left(\frac{1}{\tilde{\nu}(t)} \sum_{i=0}^{\tilde{\nu}(t)-1} \int_{\tilde{Y}_{K,i}}^{\tilde{Y}_{K,i+1}} 1_F(\tau^s\tilde{\Phi})ds \right)$$

$$= \lambda_K \int_F \left(\int_0^{Y^\wedge_{K,1}(f)} 1_F(\tau^s f)ds\right) P^o_K(df) , \qquad \text{a.e..}$$

Hence by the bounded convergence theorem

$$\lim_{t\to\infty} \frac{1}{t} \int_0^t \Pr(\tau^s\tilde{\Phi} \in F)ds = \lim_{t\to\infty} E \frac{1}{t} \int_0^t 1_F(\tau^s\tilde{\Phi})ds$$

$$= E \lim_{t\to\infty} \frac{1}{t} \int_0^t 1_F(\tau^s\tilde{\Phi})ds = \lambda_K \int_F (\int_0^{Y_{K,1}^{\wedge}(f)} 1_F(\tau^s f)ds) P_K^o(df) \ .$$

This completes the proof.

Example 3.11 (stability of the v.w.t.p. in G/G/1; FIFO queues):

Consider a G/G/1; FIFO queue. The notations and assumptions of Example 3.2 are in force with $s=1$. It was proved in Proposition 3.2 that, the r.p.@ m.p.p. (V^o,N^o), defined in Example 3.2, is a $\underline{\sigma}$-ergodic Palm version. A stationary r.p.@ m.p.p. corresponding to (V^o,N^o) we denote by (V,N). It was proved in § 2.4 (i) that the a.w.t.p. $\tilde{W}$ is stable. This yields

$$\lim_{j\to\infty} \frac{1}{j} \sum_{i=0}^{j-1} \Pr(\sigma^i(\tilde{V},N^o) \in \cdot) = \Pr((V^o,N^o) \in \cdot)$$

and hence by Theorem 3.3 we have that the r.p.@ m.p.p. $(\tilde{V},N^o)$ is stable.

Example 3.12 (stability of the v.w.t.p. in G/G/1; work conserving queue):

According to the definition, the virtual waiting time is identical for all work-conserving queues with the same generic sequence $\{T_i^o,S_i^o,K_i^o\}$. Thus the proceding example shows that the v.w.t.p. $(\tilde{V},N^o)$ is stable.

Example 3.13 (stability of the v.w.t.p. in G/G/s; FIFO queue):

Consider a G/G/s; FIFO queue. The notations and assumptions of Example 3.2 are in force. A stationary r.p.@ m.p.p. corresponding to $(\underline{V}^o,N^o)$ we denote by $(\underline{V},N)$. Was the a.w.t.p. $\{\tilde{W}_i\}$ stable with a stationary ergodic distribution then, by Theorem 3.3, the r.p. $\{\tilde{\underline{V}}\}$ would be stable. However to repeat the argument used in a single server case $(s=1)$ we should assume that

$$\Pr(\bigcup_{i=0}^{\infty} \{\underline{W}_i^o = \underline{0}\}) = 1. \tag{3.28}$$

It is known that the equation (3.28) need not be fulfilled. We refer the reader to Whitt (1972) where conditions ensuring (3.28) are given. As for as I know, there are no results allowing a proof of the stability in the general case.

Example 3.14 (stability of the queue size process in a G/G/1; FIFO queue):

In Example 3.5 we demonstrated that the queue size process $\tilde{L}$ is

$$\tilde{L}(t) = L^{\wedge}(\tilde{V},N^{o})(t) , \qquad t \in R .$$

We proved in Example 3.11 that $(\tilde{V},N^{o})$ is stable. Hence by Lemma 3.3 the r.p. $\tilde{L}$ is stable because

$$L^{\wedge}\tau^{s} = \tau_1^{s}L^{\wedge} , \qquad s \in R .$$

We also have (in view of Proposition 1.2) that the stationary queue size process L is

$$L(t) = L^{\wedge}(V,N)(t) , \qquad t \in R .$$

By Lemma 3.4, taking into account that $\tilde{L}$ is stable, we obtain

$$\lim_{t\to\infty} \frac{1}{t} \int_0^t \tilde{L}(s)ds = E\,L(o) , \qquad \text{a.e..}$$

Example 3.15 (stability of the performance r.p.):

The notions and assumptions of Example 3.7 are in force. A corresponding to (X^{o},N^{o}) stationary r.p.@ m.p.p. we denote by (X,N). Setting

$$\varphi(f) = d(0) , \qquad f = (d,n) \in F$$

in Proposition 3.4 (with $i=0$) we obtain

$$EX(0) = \frac{1}{E(T_{1,0}^{o} + T_{2,0}^{o})} E \int_0^{Y_{1,1}^{o}} X^{o}(s)ds = \frac{ET_{1,0}^{o}}{E(T_{1,0}^{o} + T_{2,0}^{o})} .$$

The quantity $\alpha = ET_{1,0}^{o}/E(T_{1,0}^{o} + T_{2,0}^{o})$ is sometimes called the stationary coefficient of availability. From (3.24) we get an intuitive interpretation of α; namely

$$\lim_{t\to\infty} \frac{1}{t} \int_0^t X^{o}(s)ds = \alpha , \qquad \text{a.e..}$$

This yields that

$$\lim_{t\to\infty} \frac{1}{t} \int_0^t EX^o(s)ds = \alpha \quad ,$$

so the limiting average availability of the system equals α.

§ 4. Relations of the first and second type.

Recall that an r.p.@ m.p.p. Φ^o is a mapping

$$\Phi^o: (A, SA, Pr) \to (F = \mathcal{D}(R,E) \times N_K, \; BF) \; .$$

It was pointed out in § 1 that $N^o = \sum_i \delta_{(Y_i^o, K_i^o)}$, where

$$\ldots < Y_{-1}^o < Y_0^o \le 0 < Y_1^o < \ldots \; .$$

Denote $\Phi^{\wedge} = (X^{\wedge}, N^{\wedge})$, where for $(d,n) \in F$

$$X^{\wedge}(d,n) = d \; , \qquad N^{\wedge}(d,n) = n \; .$$

We suppose that Φ^o is a Palm version on (A,SA,Pr). Its distribution we denote by P^o and the stationary distribution of Φ^o by P. Moreover we assume that K is a subset of Z. Let P_j^o denote the Palm distribution with respect to the mark $j \in K$. Having P we define P_j^o by (3.12). We are searching for relations between

(3.29) $\qquad v(E) = P(X^{\wedge}(0) \in E)$,

(3.30) $\qquad v_K^+(E) = Pr(X^o(0) \in E | K_0^o \in K)$,

(3.31) $\qquad v_K^-(E) = Pr(X^o(0-o) \in E | K_0^o \in K)$, $\qquad E \in BE$, $\; K \in \mathcal{K}$.

Here v is the distribution of $X(0)$, where X is a stationary r.p. corresponding to X^o. To find an interpretation of v_j^+, or v_j^-, $j \in K$, we point out that for any $i \in K$ the sequence of r.e.'s $\{X_{i,j}^+\}$ defined by

$$X^+_{i,j}(f) = X^\wedge(Y^\wedge_{i,j}(f)) , \qquad j \in Z$$

and the sequence of r.e.'s $\{X^-_{i,j}\}$ defined by

$$X^-_{i,j}(f) = X^\wedge(Y_{i,j}(f) - o) , \qquad j \in Z$$

are stationary on (F^o_i, BF^o_i, P^o_i). Moreover the distribution of $X^+_{i,j}$ is v^+_i and of $X^-_{i,j}$ is v^-_i. To prove it, it suffices to show that

$$X^+_{i,j}(f) = X^+_{i,0}(\sigma^j_i\, f)$$

and

$$X^-_{i,j}(f) = X^-_{i,0}(\sigma^j_i\, f) , \qquad j \in Z .$$

We aim to express v by means of v^-_i, v^+_i and some characteristics of Φ^o. We define these characteristics now.

Let $\{p_t(x,E),\ t \in R_+,\ x \in E,\ E \in BE\}$ be a family of stochastic kernels on (E,BE) such that for any $E \in BE$ and $t \in R_+$

$$(3.32) \qquad Pr(X^o(t) \in E | X^o(0),\ Y^o > t) = p_t(X^o(0),E) , \qquad a.e..$$

We assume that $\{p_t(x,E)\}$ may be chosen such that

(a_1) $p_0(x,E) = 1_E(x)$,

(b_1) for any fixed $E \in BE$ the function

$$E \times R_+ \ni (x,t) \to p_t(x,E) \in R_+$$

is measurable.

The family $\{p_t(x,E)\}$ describes the behaviour of the r.p. $\{X^o(t),\ t \in R\}$ within intervals $[Y^o_i, Y^o_{i+1}]$, any $i \in Z$.

Let $\{\mu^x(B),\ x \in E,\ B \in BR_+\}$ be a family of probability measures on (R_+, BR_+) fulfilling

$$(3.33) \qquad Pr(Y^o_1 \in B | X^o(0)) = \mu^{X^o(0)}(B) , \qquad a.e..$$

We assume that $\{\mu^x(B)\}$ can be choosen such that

(a_2) for any fixed $B \in BR_+$ the function

$$x \to \mu^x(B)$$

is measurable,

(b_2) for any fixed $x \in E$, $\mu^x(\cdot)$ is a probability measure on $(R_+, \mathcal{B}R_+)$. The family $\{\mu^x(B)\}$ describes the distance to the nearest point to the right of zero if $X^o(0)$ is given.

For any $K \in \mathcal{K}$, let $\{q^K(x,E)\}$ be a stochastic kernel on $(E, \mathcal{B}E)$ such that

$$\text{(3.34)} \qquad Pr(X^o(0) \in E | X^o(0-o), K_0^o \in K) = q^K(X^o(0-o), E) \;, \qquad \text{a.e..}$$

The stochastic kernel $q^K(x,E)$ describes the transition of the r.p. X^o at points from N^o.

Recall that

$$\lambda^{-1} = EY_1^o \; .$$

<u>Lemma</u> 3.5:

For any $E \in \mathcal{B}E$

$$\text{(3.35)} \qquad P(X^{\wedge}(0) \in E) = \lambda \int_{R_+} Pr(Y^o > t, X^o(t) \in E)dt \; .$$

<u>Proof</u>: The proof follows immediately from (3.8) setting

$$F = \{f = (x,n): x(0) \in E\} \; .$$

<u>Theorem</u> 3.4 (first type relation):

For any $E \in \mathcal{B}E$

$$\text{(3.36)} \qquad v(E) = \lambda \int_{E} \int_{R_+} \mu^x(t,\infty)p_t(x,E)dt \; v^+(dx).$$

<u>Proof</u>: Starting off (3.35) we have

$$v(E) = \lambda \int_{R_+} E[1_{(t,\infty)}(Y_1^o)1_E(X^o(t))]dt$$

$$= \lambda \int_{R_+} E[E(1_{(t,\infty)}(Y_1^o)1_E(X^o(t))|X^o(0), Y_1^o > t)]dt$$

$$= \lambda \int_{R_+} E\, 1_{(t,\infty)}(Y_1^o)E\, 1_E(X^o(t)|X^o(0), Y_1^o > t)dt$$

$$= \lambda \int_{R_+} E\, 1_{(t,\infty)}(Y_1^o)p_t(X^o(0),E)dt$$

$$= \lambda \int_{R_+} E\ E(1_{(t,\infty)}\{Y_1^o\} p_t(X^o(0),E)|X^o(0))dt$$

$$= \lambda \int_{R_+} E(p_t(X^o(0),E) E\ 1_{(t,\infty)}(Y_1^o)|X^o(0))dt$$

$$= \lambda \int_E \int_{R_+} \mu^x(t,\infty) p_t(x,E) dt\ v^+(dx).$$

The first type relation can be modified in the following way. Such a modification will be needed in Section 4.3. Let $i \in K$ be fixed. Instead of $\{p_t(x,E)\}$ we introduce a family $\{p_t^i(x,E), t \in R_+, x \in E, E \in \mathcal{B}E\}$ such that for any $t \in R_+$, $E \in \mathcal{B}E$

(3.37) $\quad \Pr(X^o(t) \in E|X^o(0),\ K_0^o = i,\ Y_{i,1}^o > t) = p_t^i(X^o(0),E) \quad$ a.e..

As before we suppose that $p_t^i(x,E)$ can be chosen such that

(a_3) $\quad p_0^i(x,E) = 1_E(x)$,

(b_3) for any fixed $E \in \mathcal{B}E$ the function

$$E \times R_+ \ni (x,t) \to p_t^i(x,E) \in R_+$$

is measurable.

Instead of the family $\{\mu^x(B)\}$ we need the family $\{\mu^{x,i}(B), x \in E, B \in \mathcal{B}R_+\}$ of probability measures on $(R_+, \mathcal{B}R_+)$ such that

(3.38) $\quad \Pr(Y_1^o \in B|X^o(0),\ K_0^o = i) = \mu^{X^o(0),i}(B), \quad$ a.e..

Then Theorem 3.4 can be restated in the following form.

<u>Theorem 3.4'</u>:

(3.39) $\quad v(E) = \lambda_i \int_E \int_{R_+} \mu^{x,i}(t,\infty) p_t^i(x,E) dt\ v_i^+(dx).$

We omit the proof.

Before we give the second type relation we discover a representation of v^-. We put some conditions on ϕ^o:

(3.40) $\quad \Pr(X^o(t) \in E|X^o(0),\ Y_1^o = t+s) = \Pr(X^o(t) \in E|X^o(0),\ Y_1^o > t)$,

any $E \in \mathcal{B}E,\ s,t \in R_+,$

(3.41) for any bounded $\varphi \in C(E,R)$, $x \in E$

$$\int_E \varphi(y)p_t(x,dy)$$

is a continuous function of $t \in R_+$.

Proposition 3.8:

If (3.40), (3.41) hold, then for any $E \in BE$

$$v^-(E) = \int_E \int_{R_+} p_t(x,E)\mu^x(dt)v^+(dx) . \tag{3.42}$$

Proof: We have

$$v^-(E) = Pr(X^o(0-o) \in E) = Pr(X^o(Y_1^o-o) \in E) .$$

Since X^o takes values in $D(R,E)$ we have for any bounded $\varphi \in C(E,R)$

$$\int_E \varphi(x)v^-(dx) = \int_E \varphi(x)Pr(X^o(Y_1^o-o) \in dx) \tag{3.43}$$

$$= \lim_{t\downarrow 0} \int_E \varphi(x)Pr(X^o(Y_1^o - \min(Y_1^o,t)) \in dx) .$$

Now using that

$$Pr((X^o(0), Y_1^o) \in E \times B) = \int_{E\times B} v^+(dx)\mu^x(dy)$$

we have

$$Pr(X^o(Y_1^o - \min(Y_1^o,t)) \in E)$$

$$= \int_{E\times R_+} Pr(X^o(Y_1^o - \min(Y^o,t)) \in E|X^o(0)=x, Y_1^o=s)v^+(dx)\mu^x(ds)$$

and by (3.40)

$$Pr(X^o(Y_1^o - \min(Y_1^o,t)) \in E) = \int_{E\times R_+} P_{s-\min(s,t)}(x,E)v^+(dx)\mu^x(dt). \tag{3.44}$$

Using (3.43), (3.44) we have for any bounded $\varphi \in C(E,R)$

$$\int_E \varphi(y)\Pr(X^o(Y_1^o - o) \in dy)$$

$$= \lim_{t \downarrow 0} \int_{E\times R_+} \Big(\int_E p_{s-\min(s,t)}(x,dy)\varphi(y)\Big)v^+(dx)\mu^x(ds) \ ,$$

which in turn, by (3.41), is equal to

$$\int_{E\times R_+} \Big(\int_E p_s(x,dy)\varphi(y)\Big)v^+(dx)\mu^x(ds) \ .$$

This completes the proof because φ is an arbitrary bounded function from $C(E,R)$.

Although (3.36) expresses v by v^+ and some characteristics of the process, i.e. by λ, $\{\mu^x(B)\}$, $\{p_t(x,E)\}$ it does not appear very useful. Namely it is difficult to find an explicit form of $\{\mu^x(B)\}$. However the first type relation can serve as the first step to obtaining a better relation. To state the second type relation we need the concept of the infinitesimal operator which transforms the space M of finite signed measures into itself. We endow the space M with a ω^*-topology. Let $C_K(E,R) \subset C(E,R)$ be a class of bounded functions φ with a compact support supp φ. Then for m_n, $m \in M$ we have $\omega^* - \lim m_n = m$ if and only if

(3.45) $\qquad ||m_n||$ is bounded ,

(3.46) $\qquad \lim_{n\to\infty} \int \varphi dm_n = \int \varphi dm$, $\qquad$ any $\varphi \in C_K(E,R)$.

Note that by the Stone-Weierstrass theorem (see, e.g., Rudin (1964)) to prove that $\omega^* - \lim m_n$ holds, it suffices to check (3.46) for a subset $C_K'(E,R)$ of functions such that for any compact set $E \in BE$ the family $\{f, \text{supp } f \subset E\}$ separates points of E and that there exists a function $e_E \in C_K'(E,R)$ such that $e_E(x) = 1$, any $x \in E$. To check (3.45), it suffices to apply the inequality

$$||\mu|| \le 2 \sup_{E\in BE} |\mu(E)| \ .$$

Let $\{\pi_t(x,E),\ t \in R_+,\ x \in E,\ E \in BE\}$ be a family of transition functions.

<u>Definition</u> 3.5:

The mapping

$$\mathcal{M} \ni m \to \mathcal{U}_\pi m \in \mathcal{M}$$

defined by

$$(3.47) \qquad \omega^* - \lim_{t \downarrow 0} \frac{1}{t} \left(\int_E m(dx)\pi_t(x,\cdot) - m(\cdot) \right) = \mathcal{U}_\pi m(\cdot)$$

is called the *infinitesimal operator* provided this limit exists. The set of signed measures $m \in \mathcal{M}$ for which (3.47) exists is called the domain of the operator $\mathcal{U}_\pi$ and we denote it by $\mathcal{M}_\pi$. The subscript π on the bottom of $\mathcal{U}_\pi$ or $\mathcal{M}_\pi$ denote that they are defined by the family of transition functions $\{\pi_t(x,E)\}$.

Recall that a family of transition functions $\{\pi_t(x,E)\}$ is said to be a *Markov family* if it fulfills the Chapman-Kolmogorov equation, i.e. for any $t,s \in R_+$, $x \in E$, $E \in \mathcal{B}E$

$$\pi_{t+s}(x,E) = \int_E \pi_t(x,dy)\pi_s(y,E) \ .$$

Example 3.16:

The family of transition functions of the uniform motion with velocity 1 toward the point 0 with the point 0 as an absorbing point is of the form

$$\{\pi_t(x,E) = 1_E([x-t]^+), \ t,x \in R_+, \ E \in \mathcal{B}R_+\} \ .$$

Let m be a measure on R_+ such that

$$(3.48) \qquad m(E) = \alpha\delta_o(E) + (1-\alpha) \int_E g(t)dt \ , \qquad E \in \mathcal{B}R_+ \ ,$$

where $g: R_+ \to R_+$ is bounded and measurable, and such that for constants $C_1, C_2 > 0$

$$(3.49) \qquad 0 \le g(x) < C_1 \ , \qquad x \in R_+ \ ,$$

$$(3.50) \qquad \int_0^\infty \left| \frac{g(s+t) - g(s)}{t} \right| ds < C_2 \ , \qquad t > 0 \ .$$

Let $C_K'(R_+,R_+) \subset C_K(R_+,R_+)$ be a class of differentiable functions φ such that

$$\lim_{t\downarrow 0}\sup_{x\geq 0}\left|\frac{\varphi(x-t)}{t} - \varphi'(x)\right| = 0 .$$

We have that

$$\int_0^\infty m(dx)\pi_t(x,E) - m(E)$$

$$= \int_0^\infty 1_E(s)(g(t+s)-g(s))ds + 1_E(0)\int_0^t g(s)ds .$$

Hence, by (3.49) and (3.50)

$$\sup_{E\in\mathcal{B}R_+}\left|\frac{1}{t}(\int_0^\infty m(dx)\pi_t(x,E)-m(E)\right| < C_1 + C_2 ,$$

which shows that

$$\|\frac{1}{t}(\int_0^\infty m(dx)\pi_t(x,\cdot)-m(\cdot)\| , \qquad t>0$$

is bounded. Now we have

$$\frac{1}{t}(\int_0^\infty(\int_0^\infty \varphi(y)\pi_t(x,dy))m(dx) - \int_0^\infty \varphi(x)m(dx))$$

$$= (1-\alpha)[\int_0^\infty \frac{\varphi(x-t)-\varphi(x)}{t} g(x)dx - \frac{1}{t}\int_0^t (\varphi(x-t)-\varphi(0))g(x)dx] .$$

Hence, with a view to assumptions on φ

$$\lim_{t\downarrow 0}\frac{1}{t}(\int_0^\infty(\int_0^\infty \varphi(y)p_t(x,dy))m(dx) - \int_0^\infty \varphi(x)m(dx))$$

$$= (1-\alpha)\int_0^\infty \varphi'(x)g(x)dx = (1-\alpha)g(0)\varphi(0) - \int_0^\infty \varphi(x)g(dx) .$$

Thus we worked out that for $0<a<b<\infty$

$$\mathcal{U}_\pi m(a,b) = (1-\alpha)(g(a)-g(b))$$

or in other words

$$\mathcal{U}_\pi m[x,\infty) = \frac{d}{dx} m[x,\infty) , \qquad x>0 .$$

We can generalize this example considering the uniform motion in R_+^s with the family of transition functions

$$p_t(\underline{x},E) = 1_E([\underline{x}-\underline{t}]^+) ,$$

where $\underline{x} = (x_1,\dots,x_s)$, $\underline{t} = (t,\dots,t) \in R^s$, $E \in \mathcal{B}R_+^s$. It can be proved that for $\underline{x} = (x_1,\dots,x_s)$ such that $x_i > 0$, $i=1,\dots,s$

$$\mathcal{U}_p\, m[\underline{x},\underline{\infty}) = \sum_{i=1}^{s} \frac{\partial}{\partial x_i} m[\underline{x},\underline{\infty}) .$$

Example 3.17:

Consider a Markov family of transition functions $\{p_t(x,E)\}$ on the state space $E = Z$. In such a case we write

$$p_t(i,\{j\}) = p_{ij}(t) , \qquad i,j \in Z , \quad t > 0 .$$

We assume that for a constant $C < \infty$

$$\sup_{\substack{i \in E \\ t > 0}} \frac{1}{t} |p_{ij}(t) - \delta_i(\{j\})| \le C .$$

It is known (see, e.g. Chung (1960)) that for any $i,j \in Z$

$$\lim_{t \downarrow 0} \frac{1}{t} (p_{ij}(t) - \delta_i(\{j\})) = \theta_{ij}$$

exists. Any signed measure has the form $m = \{m_i, i \in Z\}$ and we have

$$\mathcal{U}_p\, m(E) = \sum_{i \in E} m_i \theta_{ii} + \sum_i m_i \sum_{\substack{j \ne i \\ j \in E}} \theta_{ij} , \qquad E \in 2^Z .$$

Theorem 3.5 (second type relation):

Assume that (3.40), (3.41) are fulfilled and that

(3.51) $\{p_t(x,E)\}$ is a family of Markov transition functions,

(3.52) $\mu^x(\{0\}) = 0 , \qquad x \in E .$

Then $v \in \mathcal{M}_p$ and

(3.53) $\mathcal{U}_p v = \lambda(v^- - v^+) .$

Proof: From Theorem 3.4 we know that

$$v(E) = \lambda \int_E (\int_{R_+} \mu^x(t,\infty) p_t(x,E) dt) v^+(dx) .$$

First we show that

$$\sup_{E \in BE} |\frac{1}{t} (\int_E p_t(x,E) v(dx) - v(E))|$$

is bounded. We have

$$\frac{1}{t} (\int_E p_t(x,E) v(dx) - v(E))$$

$$= \frac{\lambda}{t} \int_E (\int_t^\infty (\mu^x(s-t,\infty) - \mu^x(s,\infty)) p_s(x,E) ds) v^+(dx)$$

$$- \frac{\lambda}{t} \int_0^t (\int_E \mu^x(s,\infty) p_s(x,E) v^+(dx)) ds .$$

Now

$$\frac{\lambda}{t} \int_E (\int_t^\infty (\mu^x(s-t,\infty) - \mu^x(s,\infty)) p_s(x,E) ds) v^+(dx)$$

$$= \frac{\lambda}{t} \int_E (\int_{R_+} (\int_t^\infty 1_{[s-t,s)}(u) p_s(x,E) ds) \mu^x(du)) v^+(dx) \leq \lambda$$

and

$$\frac{\lambda}{t} \int_0^t (\int_E \mu^x(s,\infty) p_s(x,E) v^+(dx)) ds \leq \lambda .$$

This demonstrates that

$$\| \frac{1}{t} (\int_E p_t(x,\cdot) v(dx) - v(\cdot)) \| \leq 4\lambda$$

Now we find $U_p v$. Let $\varphi \in C_K(E,R)$. Then

$$\frac{1}{t} (\int_E v(dx) \int_E p_t(x,dy) \varphi(y) - \int_E v(dx) \varphi(x))$$

$$= \frac{1}{t} \lambda \int_E v^+(dx) \int_t^\infty ((\mu^x(s-t,\infty) - \mu^x(s,\infty)) \int_E p_s(x,dy) \varphi(y)) ds +$$

$$- \frac{1}{t} \lambda \int_0^t (\int_E v^+(dx) \mu^x(s,\infty) \int_E p_s(x,dy) \varphi(y)) ds .$$

We let t tend to zero from the right hand side. Then

$$(3.54)\qquad \lim_{t\downarrow 0} \frac{1}{t} \lambda \int_0^t (\int_E v^+(dx)\mu^x(s,\infty) \int_E p_s(x,dy)\varphi(y))ds = \lambda \int_E v^+(dx)\varphi(x) ,$$

and

$$(3.55)\qquad \lim_{t\downarrow 0} \frac{1}{t} \lambda \int_E v^+(dx) \int_t^\infty (\mu^x(s-t,\infty) - \mu^x(s,\infty)) \int_E p_s(x,dy)\varphi(y)ds$$

$$= \lambda \int_E v^+(dx)\mu^x(ds) \int_E p_s(x,dy)\varphi(y)$$

which in turn by (3.42) is equal to

$$\lambda \int_E v^-(dx)\varphi(x) .$$

Now we verify (3.54). Since by (3.41) and (3.52) for any $x \in E$

$$\lim_{t\downarrow 0} \frac{1}{t} (\int_0^t \int_E p_s(x,dy)\varphi(y)ds - \int_0^t \mu^x(s,\infty) \int_E p_s(x,dy)\varphi(y)ds) = 0$$

we have

$$\lim_{t\downarrow 0} \frac{1}{t} \lambda \int_0^t \int_E v^+(dx)\mu^x(s,\infty) \int_E p_s(x,dy)\varphi(y)ds$$

$$= \lim_{t\downarrow 0} \lambda \int_E v^+(dx)(\frac{1}{t} \int_0^t p_s(x,dy)\varphi(y)ds) = \lambda \int_E v^+(dx)\varphi(x) .$$

To verify (3.55) we write

$$\frac{1}{t} \lambda \int_E v^+(dx) \int_t^\infty [(\mu^x(s-t,\infty) - \mu^x(s,\infty)) \int_E p_s(x,dy)\varphi(y)]dt$$

$$= \frac{1}{t} \lambda \int_E v^+(dx) \int_{R_+} \mu^x(du) \int_t^\infty 1_{[s-t,s)}(u) \int_E p_s(x,dy)\varphi(y)ds$$

which by (3.41), letting $t\downarrow 0$, tends to

$$\lambda \int_{E\times R_+} v^+(dx)\mu^x(ds) \int_E p_s(x,dy)\varphi(y) .$$

This completes the proof of Theorem 3.5.

Corollary 3.5:

$$\mathcal{U}_p v = \sum_{i \in K} \lambda_i (v_i^- - v_i^+) .$$

Proof: It suffices to show that

(3.56) $$v^+ = \sum_{i \in K} \frac{\lambda_i}{\lambda} v_i^+ ,$$

(3.57) $$v^- = \sum_{i \in K} \frac{\lambda_i}{\lambda} v_i^- .$$

Using Proposition 3.6 and 3.7 we have

$$\Pr(X^o(0) \in E) = \sum_{j \in K} \Pr(K_0^o = j) \Pr(X^o(0) \in E | K_0^o = j) = \sum_{j \in K} \frac{\lambda_j}{\lambda} v_j^+(E) ,$$

which shows (3.56). Similarly one can show (3.57).

Notice that from Corollary 3.5 using $\{q^i(x,E)\}$ we obtain

(3.58) $$\mathcal{U}_p v(\cdot) = \int_{i \in K} \lambda_i (v_i(\cdot) - \int_E v_i^-(dx) q^i(x,\cdot)) .$$

We finish up the section outlining a method of solving the Poisson type equation

(3.59) $$\mathcal{U}_p v = -m$$

with an unknown v. If v_i^-, λ_i, $\mathcal{U}_p$, $\{q^i(x,E)\}$ are known then v is a solution of (3.58) which is indeed of that type. Let $\mathcal{M}$ be the space of all signed measures with the finite variation on $(E, \mathcal{B}E)$. The space $\mathcal{M}$ with the variation norm $\| \ \|$ is a Banach space. Denote by $\mathcal{S}$ the set of invariant measures from $\mathcal{M}$ with respect to $\{p_t(x,E)\}$, i.e.

$$\mathcal{S} = \{w \in \mathcal{M}, \int_E w(dx) p_t(x,\cdot) = w(\cdot), \ t \geq 0\} .$$

Proposition 3.9:

If

(3.60) $$\int_E |\varphi(y)| \left| \int_E m(dx) \int_{R_+} p_t(x,dy) dt \right| < \infty \quad , \quad \text{any} \quad \varphi \in C_K(E,R)$$

and

(3.61) $$\lim_{t \downarrow 0} \int_E \varphi(y) p_t(x,dy) = \varphi(x) \,, \quad \text{any} \quad x \in E, \quad \varphi \in C_K(E,R)$$

then the only solution of (3.59) has the form

$$v(\cdot) = \int_{E \times R_+} m(dx)dt\; p_t(x,\cdot) + w(\cdot) \,, \qquad w \in S \,.$$

Proof: First we prove that

$$z(\cdot) = \int_{E \times R_+} m(dx)dt\; p_t(x,\cdot)$$

is a solution of (3.59). Let $\varphi \in C_K(E,R)$. Then, using (3.60) we have

$$\int_E U_p z(dx)\varphi(x) = \lim_{t \downarrow 0} \frac{1}{t} \left(\int_E z(dx) \int_E p_t(x,dy)\varphi(y) - \int_E z(dx)\varphi(x) \right)$$

$$= - \lim_{t \downarrow 0} \frac{1}{t} \int_0^t \int_E m(dx) \int_E p_s(x,dy)\varphi(y)ds$$

which, by (3.61), is equal to

$$- \int_E m(dx)\varphi(x) \,.$$

Since $U_p w = 0$ holds for any $w \in S$ we have that $z+w$ is a solution of (3.59). Now assume that there exists a solution v_1 which is not of the form $z+w$. Then clearly

(3.62) $$U_p(v_1 - v) = 0 \,.$$

However the only solution of (3.62) is an invariant measure w_1 (see e.g. Jankiewicz & Rolski (1977)) so

$$v_1 - v = w_1$$

which yields that

$$v_1 = z + (w + w_1) \,.$$

Clearly $w+w_1$ is an invariant measure which completes the proof.

Example 3.18 (Takács relation in G/GI/s; FIFO queue):

Notations and assumptions of Example 3.2 are in force. We add one more assumption that

$$(3.63)\qquad Pr(S_0^o \le t \mid \{(T_j^o, S_j^o, K_j^o),\ j \ne 0\},\ T_0^o, K_0^o) = B_{K_0^o}(t)\ , \qquad t \ge 0\ ,\ \text{a.e.}\,.$$

B_i is said to be the distribution function of the service time of customers from the i-th class. By (3.63) we have

$$(3.64)\qquad Pr(S_0^o \le t \mid \{(T_j^o, S_j^o),\ j \ne 0\},\ T_0^o) = \sum_{i \in K} \frac{\lambda_i}{\lambda} B_i(t)\ , \qquad t \ge 0\ .$$

We denote

$$B(t) = \sum_{i \in K} \frac{\lambda_i}{\lambda} B_i(t)\ , \qquad t \ge 0\ .$$

Recall the notations λ_i, $\{p_t(x,B)\}$, v, v_i^-, v_i^+ introduced in § 4. They are now applied to the r.p.@ m.p.p. $(\underline{V}^o, N^o)$. Then

$$\lambda_i^- = E(T_0^o \mid K_0^o = i)\ ,$$

$$p_t(\underline{x}, E) = 1_E([\underline{x} - \underline{t}]^+)\ , \qquad \underline{t} = (t,\dots,t) \in R_+^s\ , \qquad E \in \mathcal{B}R_+^s\ .$$

It is evident that conditions (3.40), (3.51), (3.52) are fulfilled. Note also that (3.41) holds because for any $\varphi \in C(R^s, R)$

$$(3.65)\qquad \int_{R^s} p_t(\underline{x}, d\underline{y})\varphi(\underline{y}) = \varphi([\underline{x} - \underline{t}]^+)\ .$$

It was mentioned in Example 3.16 that for any probability measure w from the domain of $\mathcal{U}_p$

$$\mathcal{U}_p w[\underline{0}, \underline{x}] = \sum_{i=1}^{s} \frac{\partial}{\partial x_i} w[\underline{0}, \underline{x}]\ , \qquad \underline{x} > \underline{0}\ .$$

Thus the second type relation is

$$(3.66)\qquad \sum_{i=1}^{s} \frac{\partial}{\partial x_i} v[\underline{0}, \underline{x}] = \sum_{j \in K} \lambda_j (v_j^-[\underline{0}, \underline{x}] - v_j^+[\underline{0}, \underline{x}])\ ,$$

where

v is the d.f. of $\underline{V}(0)$,

v_i^- is the d.f. of $\underline{M}_0$ given the condition $\{K_0^o = i\}$,
v_i^+ is the d.f. of $R(\underline{M}_0^o + \underline{S}^o)$ given the condition $\{K_0^o = i\}$,
where $\underline{S}_0^o = (S_0^o, 0, \dots, 0) \in R^s$. Recall that $R\colon R^s \to R^s$ rearranges the components of its arguments in ascending order. Note that

$$F_V(t) = v([0,t] \times R^{s-1}) \qquad (= P(V_1^\wedge(0) \le t)) , \quad t \ge 0$$

is the stationary d.f. of the virtual waiting time,

$$F_{W_i}(t) = v_i^-([0,t] \times R^{s-1}) \quad (= \Pr(M_{01} \le t)) , \quad t \ge 0$$

is the stationary d.f. of the actual waiting time of a customer from the i-th class. Let

$$F^i(t) = P(V_i^\wedge(0) \le t) , \quad t \ge 0 , \quad i=1,\dots,s .$$

Note that $F^1 = F_V$. For a d.f. H on $(R_+, \mathcal{B}R_+)$ we define $\tilde{H}$ by

$$\tilde{H}(t) = \int_0^t (1 - H(s))ds \Big/ \int_0^\infty sH(ds) .$$

The sought for Takács relation is

$$\text{(3.67)} \qquad \frac{1}{s} \sum_{i=1}^{s} F^i(t) = 1 - \rho + \sum_{i \in K} \rho_i F_{W_i} * \tilde{B}_i(t) , \qquad t \ge 0 ,$$

where

$$\rho = \frac{E\, S_0^o}{sET_0^o} , \qquad \rho_i = \frac{\lambda_i}{s} \int_{R_+} t\, B_i(dt) .$$

In a case $s=1$, (3.67) reduces to the well known

$$F_V(t) = 1 - \rho + \sum_{i \in K} \rho_i F_{W_i} * \tilde{B}_i(t) , \qquad t \ge 0 .$$

The relation (3.67) follows from the differential equation

$$\text{(3.68)} \qquad \sum_{i=1}^{s} \frac{d}{dt} F^i(t) = \sum_{i \in K} \rho_i (F_{W_i}(t) - F_{W_i} * \tilde{B}_i(t)) , \quad t \ge 0 ,$$

which we are going to establish now. Denote for a fixed $t, y \in R_+$

$$\underline{y} = (y,0,\dots,0) \in R^s ,$$

$$\underline{z}^{\oplus} = (z^{\oplus}, \dots, z_s^{\oplus}) = R(\underline{z} + \underline{y}) .$$

It holds that

$$(3.69) \qquad \sum_{i=1}^{s} 1_{[0,t]}(z_i^{\oplus}) = 1_{[0,t]}(z_1+y) + 1_{[0,t]}(z_2) + \dots + 1_{[0,t]}(z_s) .$$

Rewrite now, bearing in mind (3.64), the equation (3.66) in the form

$$(3.70) \qquad \sum_{i=1}^{s} \frac{\partial}{\partial x_i} v[\underline{0},\underline{x}] = \lambda \int_{R^s} \int_{R_+} (1_{0(\underline{x})}(z) - 1_{0(\underline{x})}(R(z+y)))B(dy)v^{-}(dz) ,$$

where

$$0(\underline{x}) = \{\underline{z}:\ 0 \le z_i \le x_i,\ i=1,\dots,s\}$$

and substitute in (3.70)

$$\underline{x}_i = (\infty,\dots,\infty,\ x_i=t,\ \infty,\dots,\infty) , \qquad i=1,\dots,s .$$

Clearly

$$\frac{\partial}{\partial x_i} v[\underline{0},\underline{x}_i] = \frac{d}{dt} F^i(t) .$$

Summing up, $i=1,\dots,s$, all results obtained, using (3.69) we get

$$\sum_{i=1}^{s} \frac{d}{dt} F^i(t) = \lambda(F_W(t) - F_W * B(t)) ,$$

where

$$F_W(t) = \Pr(M_{01} \le t) .$$

Since, by Proposition 3.6 and 3.7

$$F_W * B(t) = \sum_{i \in K} \frac{\lambda_i}{\lambda} F_{W_i} * B_i(t)$$

we have the sought for (3.68). The solution of (3.68), bearing in mind that,

$$F^i(\infty) = 1 , \qquad i=1,\dots,s$$

is indeed (3.67). In the single server case F_V is simply expressed by F_W. This seems to be impossible in a case of s-server queues; $s > 1$. Nevertheless we can express the univariate distribution of the virtual waiting time F_W by means of the first two components of $\underline{M}_0$. We have

$$(3.71) \qquad F_V(t) = \lambda \int_0^t (1 - B(t-s))\Pr(M_{01} \in dx,\ M_{02} > t) .$$

To prove (3.71), insert $\underline{x}_1$ into (3.70). Then we obtain that F_V fulfills the differential equation

$$\frac{d}{dt} F_V(t) = \lambda(G(t)-\Pr(M_{01}+S_0 \le M_{02}, M_{01}+S_o \le t) - \Pr(M_{01}+S_0 > M_{02}, M_{02} \le t)) ,$$

$$t \ge 0 ,$$

whose solution, fulfilling $F_V(\infty) = 1$, is (3.71).

Example 3.19 (second type relation in dams):

The notations and assumptions of Example 3.4 are in force. Moreover we assume that the rate of output r is a bounded function and

$$\Pr(T_0^o = 0) = 0 , \quad ET_0^o < \infty , \quad ES_0^o < \infty .$$

Let P be the stationary distribution of the Palm version (X^o,N^o) and (X,N) be the stationary r.p.@ m.p.p. corresponding to (X^o,N^o). Clearly

$$p_t(x,E) = \Pr(X^o(t) \in E \mid X^o(0), Y_1^o > t) = 1_E(z(X^o(t),t)) , \qquad t \ge 0 ,$$

$$\lambda^{-1} = ET_0^o .$$

Conditions (3.40), (3.51), (3.52) are fulfilled. It remains to check condition (3.41), namely that for any bounded $\varphi \in C(R_+,R)$, $x \in R_+$ the function

$$\int_E p_t(x,dy)\varphi(y) = \varphi(z(x,t))$$

is continuous. So (3.41) holds if for any $x \in R_+$, the function $z(x,t)$ is continuous. This is true because

$$|z(x,t'') - z(x,t')| = \int_{t'}^{t''} r(z(x,s))ds \le (\sup r)|t'' - t'| .$$

We prove now that

$$(3.72) \qquad v(x) = 1 - \alpha + \lambda \int_0^x \frac{v^-(t) - v^+(t)}{r(t)} dt , \qquad x \ge 0 ,$$

where:

v is the stationary d.f. of the content of the dam,

v^- is the stationary d.f. of the content of the dam at instants before inpouring,

v^+ is the stationary d.f. of the content of the dam at instants of inpouring. The constant α can be established from the condition $v(\infty) = 1$. For any probability measure from the domain of $\mathcal{U}_p$,

$$\mathcal{U}_p \, w[0,x] = r(x) \frac{d}{dx} w[0,x] .$$

Thus from Theorem 3.5

$$(3.73) \qquad r(x) \frac{d}{dx} v[0,x] = \lambda(v^-[0,x] - v^+[0,x])$$

which in turn implies (3.72).

The following special case is of interest if $\{T_i^o\}$ and $\{S_i^o\}$ are independent and $\{S_i^o\}$ contains i.i.d.r.v.'s with the common d.f. B. Then from (3.73) we have

$$(3.74) \qquad r(x) \frac{d}{dx} v[0,x] = \rho \tilde{b} * v^-(x) , \qquad x \ge 0 ,$$

where $\rho = \lambda E S_0^o$, $\tilde{b}(x) = (1 - B(x))/m_B$.

From (3.74) it follows that the relation for moments is

$$E[X(0)]^l = \rho E \frac{(X^o(0) + \tilde{Y}_0^o)^l}{r(X^o(0) + \tilde{Y}_0^o)} , \qquad l=0,1,...,$$

where the r.v. $\tilde{Y}_0^o$ is independent of $X^o(0)$ and is distributed according to $\tilde{B}$. Multiplying (3.74) by x^l and integrating from 0 to ∞

we obtain

$$EX^{l}(0)r(X(0)) = \rho E(X^{o}(0) + \tilde{Y}_0^{o})^{l} = \lambda \sum_{i=0}^{l} \binom{l}{i} E(X^{o}(0))^{l-i} \frac{m_{B,i+1}}{i+1} .$$

Notice that for $l = 0$,

$$E\, r(X(0)) = \rho .$$

The last equation is intuitively obvious since, under the stationary condition the average output rate, which is equal to $Er(X(0))$, must be the same as the average input rate which is equal to ρ.

<u>Example</u> 3.20 (Kuczura's model):

Consider an r.p.@ m.p.p. $\Phi^{o} = (X^{o}, N^{o})$. We assume that Φ^{o} is a Palm version. Let $E = Z$, namely X^{o} is an r.p. with a denumerable state space. Let λ_k, λ, $\{p_t(x,E)\}$, $\{q(x,E)\}$, v, v_i^- be the respective characteristics of Φ^{o} which were defined at the begining of § 4. Since $E = Z$ we denote

$$p_t(i,\{j\}) = p_{ij}(t) ,$$

$$q^{\{k\}}(i,\{j\}) = q_{ij}^{k} ,$$

$$v(\{i\}) = v_i ,$$

$$v_k^-(\{i\}) = v_{k,i}^- .$$

Suppose that $\{p_t(x,E)\}$ is a family of Markov transition functions satisfying conditions required in Example 3.17. Set $\theta_j = -\theta_{jj}$. Assuming conditions (3.40), (3.52) hold we arrive at Kuczura's relation

$$\sum_{i \neq j} v_i \theta_{ij} - v_j \theta_j = \sum_{k \in K} \lambda_k (v_{k,j}^- - \sum_i \lambda_{k,i} q_{ij}^{k}) , \quad j \in Z$$

which may be rewritten in the form

$$(3.75) \qquad \sum_{i \neq j} v_i \theta_{ij} + \sum_{k \in K} \lambda_k \sum_{i \neq j} v_{k,i}^- q_{ij}^{k}$$

$$= v_j \theta_j + \sum_{k \in K} \lambda_k v_{k,j}^- (1 - q_{jj}^{k}) , \qquad j \in Z .$$

Note that the left-hand side of (3.75) may be read as the rate of entering a state j, and the right-hand side of (3.75) may be read as the rate of leaving the state j. The form (3.75) of the second type relation (in a general case) will be the subject of interest of the next section. Consider now, in place of a general r.p.@ m.p.p., a Palm representation of the queue size process in a G/M/1; FIFO queue. Such a r.p.@ m.p.p. was defined in Example 3.5. Since the service times are independent, independent of the input and exponentially distributed with the parameter $\beta = 1/m_B$, we have that for $i,j \in N_0$

$$p_{ij}(t) = \begin{cases} e^{-\beta t}(\beta t)^i/(i-j)! \, , & j= \, ,\dots,i \\ \sum_{l=i}^{\infty} e^{-\beta t}(\beta t)^l/l! \, , & j=0 \\ 0 \, , & \text{otherwise} \, . \end{cases}$$

Thus the assumption about $\{p_t(x,E)\}$ required by Example 3.15 holds, and one can use Kuczura's relation which yields the system of equations

$$\beta v_{j+1} + \sum_{k\in K} \lambda_k \bar{v}_{k,j-1} = \beta v_j + \sum_{k\in K} \lambda_k \bar{v}_{k,j} \, , \qquad j \in N_0$$

from which we get

$$v_0 = 1 - \rho \, ,$$

$$v_j = \sum_{k\in K} \rho_k \bar{v}_{k,j-1} = \rho \sum_{k\in K} \alpha_k \bar{v}_{k,j-1} \, ,$$

where $\rho_k = \lambda_k/\lambda$, $\rho = \lambda/\beta$, $\alpha_k = \lambda_k/\lambda$; $k \in K$. Here p_j, $j \in N_0$ stand for the stationary distribution of the queue size and $\bar{v}_{k,j}$, $j \in N_0$ for the stationary distribution of the queue size just before the arrival of a customer from the k-th class.

§ 5. A rate conservative principle approach.

The main object under investigation in this section is a stationary r.p.@ m.p.p. $\Phi = (X,N)$. As in the foregoing section the space of marks K is a subset of Z. We shall also assume that the intensity λ of N is positive and finite. For convenience we suppose in this section that Φ is a canonical r.p.@ m.p.p., i.e.

$$\Phi: (F, \mathcal{B}F, P) \to (F, \mathcal{B}F) ,$$

where P is a stationary distribution on F. Exceptionally in this section we do not mark with ^ canonical r.e.'s. We aim to give another approach to the problem of relations which gives a "nice" interpretation of the second type relation in terms of a "rate conservative principle". This approach is based on the fact that the rate of entering a state (or a set) of some stationary process should be equal to the rate of leaving the state (or the set). The model presented here leads to a mathematical formalization of the rate conservative principle.

The presented approach makes it possible to treat more general cases than the one considered in Section 3.4. Namely we do not need the assumption that the family $\{p_t(x,B)\}$ is Markov. Recall that $\{p_t(x,B)\}$ was established with respect to a Palm version Φ^o. The family $\{\rho_t(x,E)\}$ introduced in this section is found with regard to the r.p. @ m.p.p. Φ. The family $\{\rho_t(x,E)\}$ takes the place of the family $\{p_t(x,E)\}$. We define $\{\rho_t(x,E)\}$ as a family of stochastic kernels on $(E, \mathcal{B}E)$ such that for any $E_1, E_2 \in \mathcal{B}E$, $t \geq 0$

$$P(X(0) \in E_1, X(t) \in E_2 | N[0,t] = 0) = \int_{E_1} \nu(dx) \rho_t(x, E_2) . \tag{3.76}$$

The disadvantage of the use of $\{\rho_t(x,E)\}$ is that we need to know the stationary version Φ of Φ^o.

The following definition precisely defines the meaning of entry and exit times.

Definition 3.6:

A point u is said to be an *entry (exit) point into (from)* $E \in \mathcal{B}E$ *of a function* $h \in \mathcal{D}(R,E)$ if there exists $\varepsilon > 0$ such that $h(u+\delta) \in E$, $h(u-\delta) \notin E$ $(h(u+\delta) \notin E,\ h(u-\delta) \in E)$ whenever $0 < \delta < \varepsilon$.

A required property is that entry and exit points by turns alternate, i.e. in the open interval between two consecutive entry (exit) points there is a unique exit (entry) point. However this is not always true. The following example, suggested by V.Schmidt, shows that some restrictions on the class of sets $E \in \mathcal{B}E$ are necessary. Let

$$E = [1,2] \cup \{\text{irrationals from } (2,\infty)\}$$

and

$$x(t) = [w_i - y_i + t]^+ , \qquad y_i \leq t < y_{i+1} ,$$

where $\dots < y_{-1} < y_0 < y_1 < \dots$ and $w_i \geq 0$, $i \in Z$. If we suppose that for every i we have $x(y_i) = w_i > 2$ and $x(y_i - o) < 1$ then in any interval $[y_i, y_{i+1})$ there is an exit point but there is no entry points.

Define for a fixed $E \in \mathcal{B}E$

$$N^*(\cdot \times \{1\}): \mathcal{D}(R,E) \to \{\textit{set of integer valued measures on } R\}$$

by

$$N^*(d)(B \times \{1\}) = \#(\{\textit{set of entry points into } E \textit{ of } d\} \cap B), \quad d \in \mathcal{D}(R,E), \; B \in \mathcal{B}R$$

and

$$N^*(\cdot \times \{0\}): \mathcal{D}(R,E) \to \{\textit{set of integer valued measures on } R\}$$

by

$$N^*(d)(B \times \{0\}) = \#(\{\textit{set of exit points into } E \textit{ of } d\} \cap B), \quad d \in \mathcal{D}(R,E), \; B \in \mathcal{B}R .$$

This is clear that N^* is not a Radon measure on $R \times \{0,1\}$ for some $d \in \mathcal{D}(R,E)$. Recall that w is a Radon measure on $R \times \{0,1\}$ if $w(B) < \infty$ for any bounded $B \in \mathcal{B}(R \times \{0,1\})$. Thus if $X = \{X(t), t \in R\}$ is an r.p. then $N^*(X)$ need not to be a p.p. because any p.p. is concentrated on $N_{\{0,1\}}$. To avoid this difficulty define $\mathcal{D}^*$ as the class of all functions $d \in \mathcal{D}(R,E)$ such that for any $a < b$, $a,b \in R$ there exists $k \in N$ such that either

$$N^*(d)([a,b] \times \{1\}) = k+1 , \qquad N^*(d)([a,b] \times \{0\}) = k$$

or

$$N^*(d)([a,b] \times \{1\}) = k+1 , \qquad N^*(d)([a,b] \times \{0\}) = k+1$$

or

$$N^*(d)([a,b] \times \{1\}) = k , \qquad N^*(d)([a,b] \times \{0\}) = k+1 .$$

Note that $d \in \mathcal{D}^*$ meets the requirement that entry and exit points by turns alternate. It turns out that $\mathcal{D}^*$ is measurable and that the mapping

$$N^*: \mathcal{D}^* \to N_{\{0,1\}}$$

is measurable. We prove these facts in an appendix attached to this section.

Now we return to the stationary r.p.@ m.p.p. $\Phi = (X,N)$. Define the class of sets $E \in E_1$ such that

$$P(\mathcal{D}^* \times N_K^{\bullet}) = 1 , \tag{3.77}$$

where

$$N_K^{\bullet} = \{n \in N_K: n(\{t\} \times K) \leq 1 , \ t \in R\}.$$

Then indeed $N^*(X)$ is an m.p.p..

All points of N we call *points of changeover*. Let $E \in E_1$ be fixed. Denote for $B \in BR$

$M_{10}^*(B)$ - the number of entry points into E of X in B which are not points of changeover,

$M_{00}^*(B)$ - the number of exit points from E of X in B which are not points of changeover,

$M_{11}^*(B)$ - the number of entry points into E of X in B which are points of changeover,

$M_{01}^*(B)$ - the number of exit points from E of X in B which are points of changeover.

We write shortly $N(t)$, $M_{ij}^*(t)$ if $B = [0,t]$ respectively. Define an m.p.p. with the space of marks $K^* = \{00,01,10,11\}$ by

$$M^*(\cdot \times \{ij\}) = M_{ij}^*(\cdot) , \qquad ij \in K^*.$$

<u>Lemma</u> 3.6:

M^* is a stationary m.p.p. such that

$$P(M^*(\{t\}) \times K^*) \leq 1, \ t \in R) = 1 .$$

The proof of Lemma 3.6 is given in the appendix attached to the section. Define

$$\mu_{ij} = E\, M_{ij}^*(1) , \qquad ij \in K^* .$$

μ_{11} (μ_{01}) is called *the rate of entering (leaving) the set* E *at points of changeover* and μ_{10} (μ_{00}) is called *the rate of entering (leaving) the set* E *not at points of changeover*. Denote

$$\mathcal{E}_2 = \{E \in \mathcal{E}_1 : \mu_{10} < \infty \; , \; \mu_{00} < \infty\} \; .$$

Proposition 3.10 (rate conservative principle):

If $E \in \mathcal{E}_2$ then

$$\mu_{10} + \mu_{11} = \mu_{00} + \mu_{01} \; . \qquad (3.78)$$

Proof: The proof follows from the inequality

$$|M^*_{10}(t) + M^*_{11}(t) - M^*_{00}(t) - M^*_{01}(t)| \le 1 \; , \qquad t > 0 \qquad \text{a.e.}$$

and from the fact that the p.p.'s N^*_{ij} are stationary.

Before we state the theorem we put more restrictions on $E \in \mathcal{BE}$. Let

$$\mathcal{E}_3 = \{E \in \mathcal{E}_1, \; P^o(\{X(0-o) \in \partial E\} \cup \{X(0) \in \partial E\}) = 0\}$$

and

$$\mathcal{E}_4 = \{E \in \mathcal{E}_2, \; P(X(0) \in \partial E) = 0\} \; ,$$

where P^o is the Palm distribution corresponding to P.

For a stationary p.p. N there are several important theorems which will be used in the proof,

(i) the theorem of Khinchin regarding the existence of the parameter

$$\lambda = \lim_{t \downarrow 0} \frac{1}{t} P(N(t) \ge 1) \; ,$$

(ii) Korolyuk's theorem which says for a stationary p.p. without multiple points

$$\lambda = E\,N(1) \; ,$$

(iii) Dobrushin's lemma stating that for a stationary p.p. without multiple points, with $\lambda < \infty$

$$P(N(t) > 1) = o(t) .$$

Recall that

$$v_i^-(E) = P_i^o(X(0-o) \in E) , \quad v_i^+(E) = P_i^o(X(0) \in E) , \quad i \in K,\ E \in \mathcal{B}\mathcal{E} ,$$

where P_i^o is the Palm distribution with respect to mark i.

Theorem 3.6:

If $E \in \mathcal{E}_3$ then

(3.79) $$\mu_{11} = \sum_{i \in K} \lambda_i \int_{E^c} v_i^-(de) q^i(e,E) ,$$

(3.80) $$\mu_{01} = \sum_{i \in K} \lambda_i \int_E v_i^-(de) q^i(e,E^c) .$$

If $E \in \mathcal{E}_4$ then

(3.81) $$\mu_{10} = \lim_{t \downarrow 0} \frac{1}{t} \int_{E^c} v(de) p_t(e,E) ,$$

(3.82) $$\mu_{00} = \lim_{t \downarrow 0} \frac{1}{t} \int_E v(de) p_t(e,E) .$$

Proof: First we prove (3.79). The proof of (3.80) is identical. We use Mecke's identity (3.11) with $((x,n = \sum_i \delta_{(y_i,k_i)}) \in \mathcal{F})$

$$\varphi((x,n),t) = \begin{cases} 1 , & t = y_i \in [0,1] \text{ and } t \text{ is an entry point into } E \text{ of } x \\ 0 , & \text{otherwise} . \end{cases}$$

Then indeed

$$\mu_{11} = \int_{\mathcal{F}} \int_{\mathbb{R}} \varphi((x,n),t) n^*(dt) P(dx \times dn)$$

and for $(x,n) \in \mathcal{F}^o$

$$\varphi(\tau^t(x,n),-t) = \begin{cases} 1 , & \text{if } 0 \text{ is an entry point into } E \text{ of } x,\ t \in [-1,0] \\ 0 , & \text{otherwise} . \end{cases}$$

Thus by (3.11)

$$\mu_{11} = \lambda P^o\ (0 \ \textit{is an entry point into}\ E\ \textit{of}\ \{X(t)\})\ .$$

Now by the total probability formula, Proposition 3.6 and 3.7

$$\mu_{11} = \sum_{i\in K} \lambda_i P_i^o\ (0\ \textit{is an entry point into}\ E\ \textit{of}\ \{X(t)\})\ ,$$

where P_i^o is the Palm distribution with respect to mark i.

If $E \in \mathcal{E}_3$ and $\lambda_i > 0$ then

$$P_i^o(X(0-o) \in \partial E) = 0\ .$$

In this case if 0 is an entry point into E then

$$X(0-o) \in \text{int}\ E\ , \qquad X(0) \in \text{int}\ E^c\ , \qquad P_i^o - \text{a.e.}\ .$$

On the other hand if

$$X(0-o) \in \text{int}\ E\ , \qquad X(0) \in \text{int}\ E^c\ , \qquad P_i^o - \text{a.e.}$$

then 0 is P_i^o - a.e. an entry point. Thus

$$\sum_{i\in K} \lambda\ P_i^o\ (0\ \textit{is an entry point into}\ E\ \textit{of}\ X) = \sum_{i\in K} \int_E v_i^-(de)q^i(e,E^c)\ .$$

The proof of (3.81) and (3.82) is more complicated.

Similarly we can show that with probability 1, any fixed point is neither an entry point nor an exit point. It follows from the stationary assumption on X and that $E \in \mathcal{E}_2$. Namely we have, for example,

$$P(0\ \textit{is an entry point}) \le E(M_{10}^*(-\varepsilon,\varepsilon) + M_{11}^*(-\varepsilon,\varepsilon)) = 2(\mu_{10} + \mu_{11})\varepsilon\ .$$

We aim now to show that

(3.83) $$\mu_{10} = \lim_{t\downarrow 0} \frac{1}{t}\ P(X(0) \notin E,\ X(t) \in E \mid N(t) = 0)\ ,$$

which can be rewritten, by (3.76), in the form $\mu_{10} = \lim_{t\downarrow 0} \frac{1}{t} \int_{E^c} v(dx)\rho_t(x,E)$.

Having established (3.81), bearing in mind that $E \in \mathcal{E}_4$, we get that if $N(t) = 0$ then

$$M^*_{10}(t) - M^*_{00}(t) = \begin{cases} -1 , & \text{if } X(0) \in E , \quad X(t) \notin E \\ 1 , & \text{if } X(0) \notin E , \quad X(t) \in E \\ 0 , & \text{otherwise,} \end{cases}$$

and hence

$$(3.84) \qquad E(M^*_{10}(t)|N(t)=0) - E(M^*_{00}(t)|N(t)=0)$$

$$= \int_{E^c} \nu(dx)\rho_t(x,E) - \int_E \nu(dx)\rho_t(x,E^c) .$$

We have, bearing in mind that M^*_{10}, M^*_{00} are stationary,

$$\mu_{i0} = \lim_{t\downarrow 0} \frac{EM^*_{i0}(t)}{t} = \lim_{t\downarrow 0} \frac{E(M^*_{i0}(t)|N(t)=0)P(N(t)=0)}{t} =$$

$$= \lim_{t\downarrow 0} \frac{E(M^*_{i0}(t)|N(t)=0)}{t} , \qquad i=0,1..$$

From (3.84) by the assumption that Φ is stationary

$$(3.85) \qquad \mu_{10} - \frac{1}{t} \int_{E^c} \nu(dx)\rho_t(x,E) = \mu_{00} - \frac{1}{t} \int_E \nu(dx)\rho_t(x,E^c) + o(1) .$$

By (3.81), the left hand side of (3.85) tends to zero as $t\downarrow 0$ which yields (3.82). Now we show (3.83). Since M^*_{10} is stationary without multiple points and $E \in \mathcal{E}_2$ we have by the theorems of Khinchin, Korolyuk and Dobrushin that

$$\mu_{10} = \lim_{t\downarrow 0} \frac{1}{t} EM^*_{10}(t) = \lim_{t\downarrow 0} \frac{1}{t} P(M^*_{10}(t)=1) .$$

From the theorem of Khinchin we have $\lim_{t\downarrow 0} P(N(t)=0)=1$ which yields

$$(3.86) \qquad \mu_{10} = \lim_{t\downarrow 0} \frac{1}{t} P(M^*_{10}(t)=1|N(t)=0)P(N(t)=0)$$

$$= \lim_{t\downarrow 0} \frac{1}{t} P(M^*_{10}(t)=1|N(t)=0) .$$

The next and last refinement of μ_{10} is

(3.87) $$\mu_{10} = \lim_{t\downarrow 0} \frac{1}{t} P(X(0) \in E,\ X(t) \notin E | N(t) = 0) .$$

To verify it we find that

(3.88) $$P([\{M^*_{10}(t) = 1\} - \{X(0) \notin E, X(t) \in E\}] | N(t) = 0) = o(t) , \quad t > 0$$

which together with (3.86) implies (3.87). So it remains show (3.88). We have

(3.89) $$P(\{M^*_{10}(t) = 1\} - \{X(0) \notin E, X(t) \in E\} | N(t) = 0)$$

$$= P(\{M^*_{10}(t) = 1\} \cap \{X(0) \notin E, X(t) \notin E\} | N(t) = 0)$$

$$+ P(\{M^*_{10}(t) = 1\} \cap \{X(0) \in E, X(t) \in E\} | N(t) = 0)$$

$$+ P(\{M^*_{10}(t) = 1\} \cap \{X(0) \in E, X(t) \notin E\} | N(t) = 0) .$$

However, bearing in mind that $E \in \mathcal{E}_4$ and 0 and t may not be either entry or exit points we have

$$P(\{M^*_{10}(t) = 1\} - \{X(0) \notin E, X(t) \notin E\} | N(t) = 0) \le$$

$$\le P(M^*_{10}(t) = 1,\ M^*_{00}(t) \ge 1 \ | N(t) = 0) \le$$

$$\le P(M^*_{10}(t) + M^*_{00}(t) \ge 2 \ | N(t) = 0) .$$

The p.p. $M^*_{10} + M^*_{00}$ is stationary without multiple points (see Lemma 3.6) with the finite intensity $\mu_{10} + \mu_{00} < \infty$. Hence by Dobrushin's lemma the above is $o(t)$. The remaining components on the right hand side of (3.89) are zero. This completes the proof of Theorem 3.6.

Theorem 3.6 and Proposition 3.10 give the explicit form of the rate conservative principle

(3.90) $$\lim_{t\downarrow 0} \frac{1}{t} \int_{E^c} v(dx)\rho_t(x,E) + \sum_{i\in K} \lambda_i \int_{E^c} v_i^-(de)q^i(e,E) =$$

$$= \lim_{t\downarrow 0} \frac{1}{t} \int_E v(dx)\rho_t(x,E^c) + \sum_{i\in K} \lambda_i \int_E v_i^-(de)q^i(e,E^c) , \qquad E \in \mathcal{E}_3 \cap \mathcal{E}_4 .$$

Corollary 3.5:

For any $E \in \mathcal{E}_3 \cap \mathcal{E}_4$

$$\text{(3.91)} \qquad \lim_{t \downarrow 0} \frac{1}{t} \left(\int_E v(de)\rho_t(e,E) - v(E) \right) = \sum_{i \in K} \lambda_i \left(v_i^-(E) - \int_E v_i^-(de) q^i(e,E) \right) .$$

Proof: The assertion of the corollary follows from (3.90) and from the identity

$$\int_{E^c} v(de)\rho_t(e,E) - \int_E v(de)\rho_t(e,E^c) = \int_E v(de)\rho_t(e,E) - v(E) .$$

Appendix:

The remaining part of the section is devoted to some technical details omitted in the considerations. Let Q denotes the set of rational numbers.

Let $E \in \mathcal{BE}$ be fixed. Define functions

$$I_1(h,t) = \begin{cases} 1 , & \text{if } t \text{ is an entry point into } E \text{ of } h \\ 0 , & \text{otherwise} \end{cases}$$

$$I_0(h,t) = \begin{cases} 1 , & \text{if } t \text{ is an exit point into } E \text{ of } h \\ 0 , & \text{otherwise .} \end{cases}$$

Lemma 3.7:

For any fixed $t \in R$, $i=0$,

$$I_i : \mathcal{D}(R,\mathcal{E}) \to R$$

is a measurable mapping.

Proof: Let $i=1$. Then

$$\{h: I_1(h,t) = 1\} = \bigcup_{i \in N} \bigcap_{\substack{a \in Q \cap (t-\frac{1}{i},t) \\ b \in Q \cap (t,t+\frac{1}{i})}} \{h,\ h(a) \in E^c,\ h(b) \in E\} .$$

Lemma 3.8:

$$\mathcal{D}^* \in \mathcal{BD}(R,\mathcal{E}) .$$

Proof: We have

$$\mathcal{D}^* = \bigcap_{\substack{a<b \\ a,b\in Q}} (\{h,\ N^*(h)([a,b]\times\{1\})=k,\ N^*(h)([a,b]\times\{0\})=k+1\}$$

$$\cup\{h,\ N^*(h)([a,b]\times\{1\})=k,\ N^*(h)([a,b]\times\{0\})=k\}$$

$$\cup\{h,\ N^*(h)([a,b]\times\{1\})=k+1,\ N^*(h)([a,b]\times\{0\})=k\}) \ .$$

Thus to demonstrate that $\mathcal{D}^* \in \mathcal{BD}(R,E)$ it suffices to show that for any $a < b$, $a,b \in Q$ and $k \in N$

$$\text{(3.92)} \qquad \{h,\ N^*(h)([a,b]\times\{i\}) = k\} \in \mathcal{BD}(R,E) \ , \qquad i = 0,1 \ .$$

For any fixed $k \in N_0$ denote

$$C_k = \{a + \frac{j}{k} : j \in N \ , \ a \le a + \frac{j}{k} \le b\} \ .$$

Then

$$\{N^*([a,b]\times\{1\}) = l\} = \liminf_{t\to\infty} \{h: \sum_{t\in C_k} I_1(h,t) = l\}$$

which by Lemma 3.7, is a measurable set. Similarly we show that

$$\{N^*([a,b]\times\{0\}) = l\} \in \mathcal{BD}(R,E) \qquad l = 0,1,\ldots$$

which completes the proof.

Lemma 3.9:

The mapping

$$N^*: \mathcal{D}^* \to \mathcal{N}_{\{0,1\}}$$

is measurable.

Proof: Sets of the form

$$G = \{n \in \mathcal{N}_{\{0,1\}},\ n(B\times\{i\}) = k\} \ , \qquad k \in N \ , \quad i=0,1$$

are generators of $\mathcal{BN}_{\{0,1\}}$. Thus it suffices to show that

$$\{h,\ N^*(h) \in G\} \in \mathcal{BD}(R,E) \ .$$

However

$$\{h,\ N^*(h) \in G\} = \{h,\ N^*(h)(B \times \{i\}) = k\}$$

which can be proved measurable similarly as in Lemma 3.8.

<u>Proof of Lemma</u> 3.6: For the proof of measurability of

$$M^*: \mathcal{D}^* \times \mathcal{N}_K^{\bullet} \to \mathcal{N}_{K^*}$$

it suffices to prove that for any $k \in K^*$,

$$M^*(\cdot \times \{k\}): \mathcal{D}^* \times \mathcal{N}_K^{\bullet} \to \mathcal{N}_{K^*}$$

is measurable. We demonstrate it in the case of $M^*(\cdot \times \{10\})$. We have $N = \sum_i \delta_{(Y_i, K_i)}$ where Y_i are r.v.'s. Then

$$\{M^*(B \times \{10\}) = k\} = \{\sum_i N^*(\{Y_i\} \times 1) 1_B(Y_i) = j\}$$

which proves the measurability because $N^*(\{Y_i\} \times 1)$ and $1_B(Y_i)$ are r.v.'s.

<u>Notes</u>.

Definitions and theorems in § 1, 2 and 3 have their counterparts in the theory of p.p.'s; see e.g. Ryll-Nardzewski (1961), Mecke (1967), Papangelou (1974), Kallenberg (1976). Theorem 3.2 is an extension and modification of the theorem of Franken & Streller (1977). Lemma 3.1 is an extension of a result from Kerstan et al (1974) or Miyazawa (1977). Similar results to the result of Theorem 3.5 were obtained by Zähle (1980). A construction of a stationary v.w.t.p. in single server queues was given by Kalähne (1976), however her construction differs from the one presented here. The idea of using the theory of stationary m.p.p.'s to demonstrate Little's formula is due to Franken (1976). Franken (1978) proved that in stationary failure r.p.'s X, the coefficient of availability equals EX(0). The relations of the first and second type were established in this form by Jankiewicz & Rolski (1977). The methods used in § 4 are extensions of ones from Rolski (1977a). However in both the papers mentioned so called piecewise Markov processes were considered which one could simply extend to a Markov process. The first and second Takács relations in GI/GI/1 queues were given by Takács (1955) and (1963) respectively. The second Takács relation attracted attention especially. In the G/G/1 case it was established in a slightly different form by Loynes (1962). In G/G/k queues it was found by Miyazawa (1976a) (formula (3.67)) and by Kopociński & Rolski (1977) (formula (3.71)). The relation of the first and second type in a case of a discrete state space were first given by Kuczura (1973). For the rate conservative principle approach to the relation of the second type see König, Rolski, Schmidt & Stoyan (1978), Rolski (1978) and König & Schmidt (1980a), (1980b). We also mention the paper of Krakowski (1973) where the idea of the rate conservative principle was used heuristicaly in queueing theory and the paper of Brill and Posner (1977). In the latter paper some interesting applications of the rate conservative principle to single server queues are given. Schmidt (1978) applied the results of König et al (1978) to investigation of some queueing systems.

CHAPTER 4. MISCELLANEOUS EXAMPLES

§ 1. Inequalities and identities.

Consider an r.p.@ p.p. $\Phi^o = (X^o, N^o)$ on (A, SA, Pr). Assume that Φ^o is a Palm version. In this section we try to answer the question of when $v = v^-$ or $v \le_{st} v^-$ or $v^- \le_{st} v$. We shall recall the notion of stochastic ordering $(\le_{st})$ later in Definition 4.1. Having established a relation between v and v^- we can find either an equation determining v^- or an operator inequality which sometimes makes it possible to find useful bounds for v or v^-. For example if $v = v^-$ then from the second type relation we obtain that v fulfills

$$(4.1) \qquad \mathcal{U}_p v(\cdot) = \lambda\left(v(\cdot) - \int_E q(x,\cdot)v(dx)\right) .$$

Recall the notations λ, $\{\mu^x(B)\}$, $\{p_t(x,E)\}$ defined in § 3.4. We assume that Φ^o is such that (3.40) and (3.41) from § 3.4 hold. Then by Proposition 3.8

$$v^-(\cdot) = \int_E \int_{R_+} p_t(x,E)\mu^x(dt)v^+(dx) .$$

The next proposition gives a condition for $v = v^-$.

Proposition 4.1:

If conditions (3.40), (3.41) from § 3.4 hold and

$$\mu^x[0,t] = 1 - e^{-\lambda t} , \qquad t \ge 0$$

then

$$v = v^- .$$

Proof: Using the first type relation we have

$$v(E) = \int_E \int_{R_+} \lambda e^{-\lambda t} p_t(x,E)dt\, v^+(dx) = v^-(E) .$$

We assume now that the r.p. $\{X^o(t)\}$ takes values in R^l i.e. $E = R^l$.

Definition 4.1:

Let n, m be two probability measures on (R^l, BR^l). It is said that n is *stochastically smaller than* m $(n \le_{st} m)$ if for any non-decreasing bounded below function h

$$\int_{R^l} h\,dn \le \int_{R^l} h\,dm .$$

In the univariate case $(E = R)$

$$n \le_{st} m$$

is equivalent to

$$n(x,\infty) \le m(x,\infty) , \qquad \text{any } x \in R .$$

This is not generally true for $E = R^l$, $l \ge 2$.

Proposition 4.2:

Assume that $\mu^x(\cdot) = \mu(\cdot)$, $x \in E$. If

(i) $\tilde{\mu} \le_{st} \mu$ (or $\mu \le_{st} \tilde{\mu}$);

(ii) for any fixed $\underline{a}$, $\underline{x} \in R^l$ the function

$$t \to p_t(\underline{x},(\underline{a},\underline{\infty}))$$

is nonincreasing then

$$v^- \le_{st} v \qquad (\text{or } v \le_{st} v^-) .$$

Proof: Using the first type relation, for any $\underline{a} \in R^l$

$$v(\underline{a},\underline{\infty}) = \lambda \int_{R^l} \int_{R_+} \mu(t,\infty) p_t(\underline{x},(\underline{a},\underline{\infty}))dt\, v^+(dx)$$

$$= \int_{R^l} \int_{R_+} p_t(\underline{x},(\underline{a},\underline{\infty}))\tilde{\mu}(dt)v^+(dx) \ge$$

$$\ge \int_{R^l} \int_{R_+} p_t(\underline{x},(\underline{a},\underline{\infty}))\mu(dt)v^+(d\underline{x}) = v^-(\underline{a},\underline{\infty}) .$$

Giving up some generality in the last proposition, namely assuming that $E = R$ we are able to consider a case in which $\mu^x(B)$ does depend on $x \in E$.

Proposition 4.3:

Let $E = R$ and

(i) $\tilde{\mu}^x \leq_{st} \mu^x$, $x \in R$,

(ii) for any fixed $a, x \in R$ the function

$$t \to p_t(x,(a,\infty))$$

is nonincreasing,

(iii) the functions

$$m^x = \int_{R_+} t\mu^x(dt) \quad \text{and} \quad \int_{R_+} p_t(x,(a,\infty))\mu^x(dt)$$

are simultaneously increasing (decreasing) functions of x. Then

$$v^- \leq_{st} v .$$

Proof: Using the first type relation, for any $a \in R$

$$v(a,\infty) = \lambda \int_R \int_{R_+} \mu^x(t,\infty)p_t(x,(a,\infty))dtv^+(dx) \geq$$

$$\geq \lambda \int_R m^x \int_{R_+} p_t(x,(a,\infty))\tilde{\mu}^x(dt)v^+(dx) .$$

Now we note that if $g,h: R \to R$ are two simultaneously increasing (decreasing) functions then for any r.v. X

$$Eg(X)h(X) \geq Eg(X)Eh(X)$$

provided the expected values exist. Hence

$$\lambda \int_R m^x \int_{R_+} p_t(x,(a,\infty))\mu^x(dt)v^+(dx) \geq$$

$$\geq (\lambda \int_R m^x\, v^+(dx))(\int_R \int_{R_+} p_t(x,(a,\infty))\mu^x(dt)v^+(dx)) = v^-(a,\infty) .$$

Example 4.1:

Consider a GI/GI/s; FIFO queue. Notations and assumptions of Example 3.18 are in force. Moreover we assume that

$$\frac{1}{ET_0} \int_x^\infty \Pr(T_0^o > t)dt \le \Pr(T_0^o > x) , \qquad x \ge 0. \tag{4.2}$$

In reliability theory the distribution of T_0^o fulfilling (4.2) is said to have the "New Better than Used in Expectation" property. Denote

$$B(t) = \Pr(S_0^o \le t) , \qquad t \ge 0.$$

In Example 3.18 we noticed that

$$P_t(\underline{x},(\underline{a},\infty)) = 1_{(\underline{a},\infty)}([\underline{x} - \underline{t}]^+) .$$

We also need to find the family $\{\mu^x(B)\}$. Due to the assumption that $\{T_i^o\}$ consists of i.i.d.r.v.'s and that sequences $\{T_i^o\}$ and $\{S_i^o\}$ are independent we have $\mu^x[0,t] = \Pr(T_0^o \le t)$, $t \ge 0$. Thus clearly conditions (i) and (ii) of Proposition 4.2 are fulfilled. Hence

$$F_V(t) \le F_W(t) , \qquad t \in R . \tag{4.3}$$

Since $V_1(0) \le \dots \le V_s(0)$ we have, using (4.3)

$$F_W(t) \ge F^1(t) \ge \dots \ge F^s(t) , \qquad t \ge 0 ,$$

and hence from (3.67)

$$F_W(t) \ge 1 - \rho + \rho F_W * \tilde{B}(t) , \qquad t \ge 0 . \tag{4.4}$$

Repeated substitution for F_W on the right-hand side of (4.4) yields the following bound

$$F_W(t) \ge \sum_{j=0}^{\infty} (1-\rho)\rho^j \tilde{B}^{*j}(t) , \qquad t \ge 0 . \tag{4.5}$$

In the right-hand side of (4.5) we recognize the stationary actual waiting time in an M/GI/1 queue with the same distribution of the service time and the arrival intensity λ/s. Thus the stationary actual waiting time in a GI/GI/s; FIFO queue with the inter-arrival time fulfilling

(4.2), is stochastically smaller than the stationary actual waiting time in an M/GI/1 queue with the same distribution of the service time and the arrival intensitiy λ/s. In the case $s = 1$ by Proposition 4.1, we have in (4.4) a case of equality if

$$\Pr(T_0^o \le t) = 1 - \exp(-\lambda t) , \qquad t \ge 0 .$$

So we have equality in (4.5) and we then obtain the well known formula for the stationary actual waiting time in an M/GI/1 queue.

<u>Example</u> 4.2 (identity and inequality in the theory of dams):

Notations and assumptions of Example 3.4 and 3.19 are in force. Moreover we assume that $\{T_i^o\}$ and $\{S_i^o\}$ are independent, $\{S_i^o\}$ contains i.i.d.r.v.'s and $\{T_i^o\}$ contains i.i.d.r.v.'s with the common d.f.

$$A(t) = 1 - \exp(-\lambda t) , \qquad t \ge 0 . \tag{4.6}$$

Then the density $h(t)$ of the stationary content of the dam defined by

$$v[0,t] = 1 - \gamma + \int_0^t h(t)dt , \qquad t \ge 0 ,$$

is the solution of the following integral equation

$$r(t)h(t) = \rho((1-\gamma)\tilde{b}(t) + \int_0^t \tilde{b}(t-s)h(s)ds) , \qquad t \ge 0 . \tag{4.7}$$

To prove it we find that by (4.6) we have

$$v = v^-$$

so from (3.74) we have (4.7). Now assume that the d.f. A of T_0^o fulfills

$$\tilde{A} \le_{st} A .$$

The condition (ii) of Proposition 4.2 holds because the process between two consecutive instants of inpouring is nonincreasing. Thus $v^- \le_{st} v$. Hence from (3.74)

$$\int_0^t r(s)h(s)ds = \rho v^- * \tilde{B}(t) \ge \rho v * \tilde{B}(t) , \qquad t \ge 0 .$$

Hence the sought for integral inequality for the density h is

(4.8) $$\int_0^t r(s)h(s)ds \geq \rho((1-\gamma)\widetilde{B}(t) + \int_0^t \widetilde{B}(t-s)h(s)ds) , \qquad t \geq 0 .$$

§ 2. Kopocińska's model.

In this section we apply the results of § 3.4 to the problem which was set up by Kopocińska (1977b). The problem was to find the relation between the joint stationary distribution of an r.p. at zero and the mark of the last point before zero and the distributions v_k^-, v_k^+. For example analysing a queueing priority system we want to know the joint distribution of the queue size and the priority of the last arriving customer.

Consider a stationary r.p.@ m.p.p. $\Phi^o = (X^o, N^o)$ on (A, SA, Pr). We assume that Φ^o is a Palm version with respect to a mark k. In the section, k denotes a fixed mark. We assume that the space of marks K is a subset of Z. Define the r.p. $\{K_k^o(t),\ t \in R\}$ by

$$K_k^o(t) = \begin{cases} 1 , & \text{if the last point before } t \text{ has the mark } k , \\ 0 , & \text{otherwise} . \end{cases}$$

Trajectories of K^o are from $\mathcal{D}(R, R_+)$. We shall study the r.p. (X^o, K_k^o) associated with the m.p.p. N^o. Denote $\Phi_k = ((X^o, K_k^o), N^o)$. Since Φ^o is a Palm version we have that Φ_k^o is. It follows from Proposition 1.2 if we note that

$$\Phi_k^o = \Phi_k^o(\Phi^o)$$

and that

$$\sigma_k^i \Phi_k^o(\Phi^o) = \Phi_k^o(\sigma_k^i \Phi^o) .$$

Let $\Phi_k = ((X, K_k), N)$ be a stationary r.p.@ m.p.p. corresponding to Φ_k^o. It exists by considerations of Section 3.2 provided

$$EY_{k,1}^o < \infty .$$

In this section we find the first and second type relation for Φ_k^o. The relations will be expressed by characteristics $\{p_t(x, B)\}$,

$\{\mu^{x,k}(B)\}$, λ_k defined before Theorem 3.4! Denote

$$v_k(E) = Pr(X(0) \in E, K_k(0) = 1) .$$

Applying Theorem 3.4' to Φ_k^o we obtain the first type relation

(4.9) $$v_k(E) = \lambda_k \int_E \int_{R_+} \mu^{x,k}(t,\infty) p_t^k(x,E) dt\, v_k^+(dx) .$$

To obtain the second type relation we must impose some conditions. Recall that K_0^o denotes the mark of a point at zero. Define

(4.10) for any $E \in \mathcal{B}\mathcal{E}$, $s,t \geq 0$,

$$Pr(X^o(t) \in E | X^o(0), K_0^o = k, Y_{k,1}^o = t+s)$$

$$= Pr(X^o(t) \in E | X^o(0), K_0^o = k, Y_{k,1}^o > t) ,$$

(4.11) for any bounded $\varphi \in C(\mathcal{E},R)$, $x \in \mathcal{E}$ the function

$$\int_{\mathcal{E}} p_t(x,dy)\varphi(y)$$

is continuous in $t \geq 0$,

(4.12) $\{p_t^k(x,B)\}$ is a family of Markov transition functions,

(4.13) $$Pr(K_i^o = k+1 | K_{i-1}^o = k) = 1,$$

(4.14) $$\mu^{x,k}(\{0\}) = 0 , \qquad x \in \mathcal{E} .$$

In the next proposition we find a relation between v_{k+1}^- and v_k^+.

<u>Proposition</u> 4.4:

If (4.10) - (4.13) are fulfilled, then

$$v_{k+1}^-(E) = \int_{\mathcal{E}} \int_{R_+} p_s^k(x,E)\mu^{x,k}(ds) v_k^+(dx)$$

<u>Proof</u>: By the definition

$$v_{k+1}^{-}(E) = \frac{\Pr(X^{o}(Y_{k,1}^{o} - o) \in E,\ K_0^{o} = k)\ \Pr(K_0^{o} = k)}{\Pr(K_0^{o} = k)\ \Pr(K_0^{o} = k+1)}$$

which by Proposition 3.6 and 3.7 yields

$$(4.15) \qquad v_{k+1}^{-}(E) = \frac{\lambda_k}{\lambda_{k+1}} \Pr(X_0^{o}(Y_{k,1}^{o} - o) \in E \mid K_0^{o} = k) \ .$$

Substituting $E = \mathcal{E}$ in (4.15) we obtain $\lambda_k = \lambda_{k+1}$. As in Proposition 3.8 we can show that

$$\Pr(X^{o}(Y_{k,1}^{o} - o) \in E \mid K_0^{o} = k) = \int_{\mathcal{E}} \int_{R_+} p_s^k(x,E)\mu^{x,k}(ds)v_k^{+}(dx) \ .$$

<u>Proposition</u> 4.5:

If (4.10) - (4.14) are fulfilled, then

$$\mathcal{U}_{p^k} v_k = \lambda_k (v_{k+1}^{-} - v_k^{+}) \ .$$

<u>Proof</u>: The proof proceeds as the proof of Theorem 3.5 using Proposition 4.4 and the first type relation (4.9).

§ 3. <u>Equivalence of distributions of embedded chains in the queue size process</u>.

In this section we study a general problem without assumptions on a queueing structure. Namely, we find that under fairly general conditions the distribution of the queue size at arrival instants equals the distribution of the queue size at departure instants.

Consider a queueing system with a generic sequence $\{T_i^o, S_i^o, K_i^o\}$. As usual we suppose that the generic sequence is metrically-transitive. Notice that nothing is supposed on a queueing structure. We only assume that there exists a finite stationary a.w.t.p. $\{W_i^o\}$ such that $\{W_i^o, T_i^o, S_i^o, K_i^o\}$ is metrically-transitive. For example in G/**G**/1; FIFO queues such sequence (assuming a stability condition) is ensured by Loynes' lemma. In more general G/G/1 queues with work-conserving normal disciplines, such a sequence is ensured by Theorem 2.5. We are going to show the equivalence of distributions of embedded chains in the queue size process at instants prior to arrivals or at instants after departures. This will be done by coming out of a Palm representation of the queue size process. It was defined in Example 3.6 that

$$L^o(t) = \#\{j:\ Y_j^o \le t < Y_j^o + W_j^o + S_j^o\} , \qquad t \in R .$$

The associated m.p.p. was defined as

$$N^o = \sum_j \delta_{(Y_j^o, K_j^o)} .$$

Recall that the r.p.@ m.p.p. (L^o, N^o) is a $\underline{\sigma}$-ergodic Palm version. Embedded queue size chains we define by: just prior to arrival instants

$$L_i^- = L^o(Y_i^o - o) , \qquad i \in Z ,$$

just after departure instants

$$L_i^+ = L^o(Y_i^o + W_i^o + S_i^o) , \qquad i \in Z .$$

It turns out that (L^-, L^+) is metrically-transitive. To prove it, consider $n = \sum_i \delta_{(y_i, [v_i, k_i])} \in N_{R_+ \times K}$. Define the mapping $l^-: N_{R_+\times K} \to R^Z$ by

$$l_j^-(n) = \sum_i 1_{[y_i, y_i + v_i)}(y_j - o) , \qquad j \in Z ,$$

and the mapping $l^+: N_{R_+\times K} \to R^Z$ by

$$l_j^+(n) = \sum_i 1_{[y_i, y_i + v_i]}(y_j + v_j) , \qquad j \in Z .$$

The mapping $(l^-, l^+): N_{R_+\times K} \to (R\times R)^Z$ is measurable but we omit the proof of this fact. Recall the mapping σ on $N_{R_+\times K}$ defined in § 3.1. Namely $\sigma n = \sum_i \delta_{(y_i - y_1, [v_i, k_i])}$. On $(R\times R)^Z$ we have the mapping θ defined by $\theta\{x_i, y_i\} = \{x_{i+1}, y_{i+1}\}$. It is easy to see that $(l^-\sigma, l^+\sigma) = \theta(l^-, l^+)$. Define

$$\bar{N}^o = \sum_j \delta_{(Y_j^o, [V_j^o, K_j^o])} .$$

Clearly $\bar{N}^o$ is a $\underline{\sigma}$-ergodic Palm version and

$$(L^-,L^+) = (l^-(\bar{N}^o),(l^+(\bar{N}^o)).$$

Thus, by Proposition 1.2, (L^-,L^+) is a metrically-transitive sequence. Define

$$C_i^o = \min\{j > i:\ W_i^o = 0\} - \max\{j \le i:\ W_i^o = 0\} + 1\ , \qquad i \in Z.$$

In other words C_i^o is the number of served customers in the busy cycle to which the i-th customer belongs. It can be proved that the sequence $\{C_i^o\}$ is metrically-transitive.

Theorem 4.1:

If $EC_0^o < \infty$ then for any Borel bounded function ψ: $R \to R$

$$\lim_{j\to\infty} \frac{1}{j} \sum_{i=0}^{j-1} \psi(L_i^-) = \lim_{j\to\infty} \frac{1}{j} \sum_{i=0}^{j-1} \psi(L_i^+)\ , \qquad \text{a.e. .}$$

Proof: The proof is based on the fact that

$$\sum \psi(L_i^-) = \sum \psi(L_i^+)\ , \tag{4.16}$$

where the summation is taken over indices of all customers in a busy cycle. To demonstrate this consider the set of all indices of customers in a busy cycle. Then to any i from this set there exists j also from this set such that $L_i^- = L_j^+$ and the correspondence is one-to-one. From (4.16) it follows

$$\left|\frac{1}{j} \sum_{i=0}^{j-1} \psi(L_i^-) - \frac{1}{j} \sum_{i=0}^{j-1} \psi(L_i^+)\right| \le 2 \sup_x |\psi(x)|\ \frac{C_{j-1}^o + C_0^o}{j}\ , \qquad j \in N. \tag{4.17}$$

Since (L^-,L^+) is stationary we have that

$$\lim_{j\to\infty} \frac{1}{j} \sum_{i=0}^{j-1} \psi(L_i^-)\ , \qquad \lim_{j\to\infty} \frac{1}{j} \sum_{i=0}^{j-1} \psi(L_i^+)\ , \qquad \text{a.e.}$$

exist. Thus, by (4.17) it suffices to prove

$$\lim_{j\to\infty} \frac{C_j^o}{j} = 0 \qquad \text{a.e. .} \tag{4.18}$$

However (4.18) follows because $\{C_i^o\}$ is a metrically-transitive sequence and $EC_0^o < \infty$.

Corollary 4.1:

Stationary d.f.'s of $\{L_i^-\}$ and $\{L_i^+\}$ are identical.

A similar result can be obtained for distributions of embedded chains in the queue size process of customers from a given class $k \in K$. Let $k \in K$ be fixed and denote by

Y_{kj}^o - the arrival instants of the j-th customer in the k-th class,

S_{kj}^o - the service time of the j-th customer in the k-th class.

We assume that there exists a finite a.w.t.p. $\{W_{kj}^o\}$ such that $\{W_{kj}^o, T_{kj}^o = Y_{k,j+1}^o - Y_{kj}^o, S_{kj}^o\}$ is metrically-transitive.

Define the queue size process of customers from the k-th class by

$$L_k^o(t) = \#\{j: Y_{kj}^o \le t < Y_{kj}^o + W_{kj}^o + S_{kj}^o\},$$

and the chains embedded in it by

$$L_{kj}^- = L_k^o(Y_{kj}^o - o) ,$$

$$L_{kj}^+ = L_k^o(Y_{kj}^o + W_{kj}^o + S_{kj}^o) , \qquad j \in Z .$$

To L_k^o we associate the p.p.

$$N_k^o = \sum_j \delta_{Y_{kj}^o} .$$

The r.p.@ p.p. (L_k^o, N_k^o) is a Palm version. Let C_{k0}^o denote the number of served customers in the busy cycle of customers from the k-th class to which customer 0 from the k class belongs, i.e.

$$C_{ki}^o = \min\{j > i: W_{kj}^o = 0\} - \max\{j \le i: W_{kj}^o = 0\} + 1 \qquad i \in Z .$$

Theorem 4.2:

If $EC_{k0}^o < \infty$ then for any Borel function $\psi: R_+ \to R_+$

$$\lim_{j\to\infty} \frac{1}{j} \sum_{i=0}^{j-1} \psi(L_{ki}^-) = \lim_{j\to\infty} \frac{1}{j} \sum_{i=0}^{j-1} \psi(L_{ki}^+) , \qquad \text{a.e. .}$$

Corollary 4.2:

Stationary d.f.'s of $\{L_{k,i}^-\}$ and $\{L_{k,i}^+\}$ are identical.

Notes

It was first discovered by Mori (1975) that in a GI/GI/s; FIFO queue, under an assumption on the input process, the stationary d.f. of the virtual waiting time is stochastically greater than the stationary d.f. of the actual waiting time. This phenomena has been recently studied by König & Schmidt (1980c). The fact that the stationary d.f. of the actual waiting time in a GI/GI/s queue is stochastically smaller than the stationary d.f. of the actual waiting time in the respective M/GI/1; FIFO queue was discovered by Rolski (1979). Independently this fact was noticed by Miyazawa (1976b) and Stoyan (1977) however they did it for single server queues only. The model of a dam with a Poisson input was considered among others by Çinlar & Pinsky (1972) and Harrison & Resnick (1976). The integral equation given in (4.7) belongs to Harrison & Resnick (1976).

The relation studied in § 2 was also investigated by Jankiewicz (1979a), (1979b). However Kopocińska and Jankiewicz considered an r.p. which can simply be extended to a Markov process.

For a few particular queues the fact that stationary d.f.'s of $\{L_i^-\}$ and $\{L_i^+\}$ are equal is well known. Usually this result is proved by calculating out each of these distributions and then varifying that they are equal, see e.g. Cohen (1969). The approach used in § 3 is similar to one given by Franken (1976) and Klimov (1979).

Chapter 5. Application to single server queues

§ 1. Introductory remarks.

This chapter is a contribution to the theory of single server queues. We are not going to give an exhaustive account of such queues. We are only going to demonstrate how the theory developed can help in dealing with some non-standard queues. In § 2 and 3 we study single server queues with FIFO disciplines not defined by means of a standard stationary generic sequence. In § 4 we study G/G/1 queues with work-conserving, normal disciplines. We demonstrate there how one may resolve such models into standard forms. In § 5 and 6 we study G/G/1; FIFO and GI/GI/1; FIFO queues.

We would like to direct our attention to a typical scheme employed herein. Considering an r.p. describing an interesting queueing characteristics, by appriopriate association with an m.p.p. we find a respective Palm representation of the characteristic. Usually we can quess at the form of the looked for Palm representation, and then verify that the r.p. under investigation is equivalent to this Palm representation. In some cases (e.g. § 2) Lemma 1.2 is useful for proving equivalence. Having established a form of the Palm representation by applying results of Chapter 3 we obtain the Takács relations and the Little formula.

Within a section all r.e.'s are defined on a common probability space $(A, SA, \Pr)$. Such a space always exists. For example, deriving an r.p.@ m.p.p. $\tilde{\Phi}$, a corresponding Palm version Φ^{o} and stationary r.p. @ m.p.p. Φ with distributions $\tilde{P}$, P^{o} and P respectively we may take

$$A = F \times F \times F \;, \qquad SA = B(F \times F \times F) \;, \qquad \Pr = \tilde{P} \otimes P^{o} \otimes P$$

and for $\underline{f} = (f_1, f_2, f_3) \in A$ set

$$\tilde{\Phi}(\underline{f}) = f_1 \;, \qquad \Phi^{o}(\underline{f}) = f_2 \;, \qquad \Phi(\underline{f}) = f_3 \;.$$

Recall that for a d.f. F we write

$$\tilde{F}(x) = \int_0^x (1-F(t))dt / \int_0^\infty (1-F(t))dt \quad .$$

§ 2. Single server queue with periodic input.

Consider a single server, infinite capacity queue at which customers arrive according to a periodic p.p.. In some cases a periodic input p.p. naturally arises. For example considering a queue over a period of a few months we can find that the arrival stream is periodic and the period is equal to 24 hours.

We start with introducing the notion of a periodic p.p.. For simplicity we assume that the period is equal to 1 . Let $\{H_i,\ i \in Z\}$ be a sequence of p.p.'s without multiple points. We assume that any p.p. H_i has points in $[0,1)$. One might say that points of H_i are instants of arrival during the i-th day (a day is supposed to be a unit interval). We assume that

$$0 < EH_o[0,1) = \lambda < \infty \ ,$$

and that the sequence $\{H_i\}$ is metrically-transitive. Define the following m.p.p. on R with a two element space of marks $K = \{0,1\}$ by

$$N_1^o(B \times \{0\}) = \sum_j H_j(B \cap [j,j+1) - j) \ ,$$

$$N_1^o(B \times \{1\}) = \#\{i:\ i \in B \cap Z\} \ .$$

The m.p.p. N_1^o is a Palm version with respect to the mark 1 . Since $\lambda_1 = 1 < \infty$, we have the existence of a stationary m.p.p. N corresponding to N_1^o. Note that

$$\lambda_0 = EN([0,1) \times \{0\}) = \lambda > 0 \ .$$

Hence there exists a Palm version of N with respect to the mark 0 . Denote it by N_0^o. We denote the consecutive coordinates of points with the mark 0 of N_0^o by $Y_{0,i}^o$, $i \in Z$ and of N_1^o by $Z_{0,i}^o$, $i \in Z$. Set

$$T_{0,i}^o = Y_{0,i+1}^o - Y_{0,i}^o \ , \qquad i \in Z \ .$$

Since $\{H_i\}$ is metrically-transitive we have that the m.p.p. N_1^o is $\underline{\sigma}_1$-ergodic. Thus by Lemma 3.2, N is metrically-transitive and hence N_0^o is $\underline{\sigma}_0$-ergodic. This yields that

$$\{T_{0,i}^o = Y_{0,i+1}^o - Y_{0,i}^o \ , \quad i \in Z\}$$

is metrically-transitive. Moreover it can be shown that

$$ET_{0,i}^o = \lambda^{-1} \ ,$$

because λ is the intensity of points with the mark zero.

Note two important examples of a periodic p.p.. If $\{H_i, \ i \in Z\}$ is a sequence of i.i.d.p.p.'s and H_0 is a Poisson p.p. on $[0,1)$ with the intensity measure Λ then N_1^o is a Poisson p.p. on R with the intensity measure

$$\sum_{i=-\infty}^{\infty} \Lambda(\cdot \cap [i,i+1) - i) \ .$$

In the second example $H_i = H_0, \ i \in Z$ and H_0 is a finite, deterministic system of points in the interval $[0,1)$. Then indeed N_1^o is deterministic, but both N and N_0^o are random m.p.p.'s.

Consider now a single server, infinite capacity queue, with the FIFO discipline. Customers arrive at the system according to points of N_1^o with the mark 0. The service times $S_i^o, \ i \in Z$ are independent and independent on the input process, identically distributed with a common d.f. B. We assume an initial condition that the queue is empty before zero. The v.w.t.p. in the queue we denote by $\tilde{V} = \{\tilde{V}(t), \ t \in R\}$. Other r.p.'s of interest are:

$\tilde{V}^s = \{\tilde{V}(i+s), \ i \in Z\}$ - the work-load process at a given instant s of a day $(s \in [0,1)$ is fixed),

$\tilde{W} = \{\tilde{W}_i = \tilde{V}(Z_{0,i}^o - o), \ i \in Z\}$ - the a.w.t.p.,

$\tilde{L} = \{\tilde{L}(t) = L^{\wedge}(\tilde{V},\tilde{N}_1^o)(t), \ t \in R\}$ - the queue size process ($L^{\wedge}$ was defined in Example 3.5 and $\tilde{N}_1^o = \sum_i \delta_{Z_{0,i}^o}$) .

<u>Proposition</u> 5.1:

If $\rho = \lambda ES_0^o < 1$ then the r.p.'s $\tilde{V}, \ \tilde{W}, \ \tilde{L}$ are stable. Moreover the r.p. $\tilde{V}^s$ is strongly stable.

<u>Proof</u>: We use the coupling method. Let $(V_0^o, \tilde{N}_0^o = \sum_i \delta_{Y_{0,i}^o})$ be the v.w.t.p.@ p.p. defined by the generic sequence $\{T_{0,i}^o, S_i^o\}$ according

to the procedure given in Example 3.2 (a Palm representation of the v.w.t.p.). Instead of $(V_0^o, \tilde{N}_0^o)$ we may consider the r.p.@ m.p.p. (V_0^o, N_0^o). The r.p.@ m.p.p. (V_0^o, N_0^o) is a Palm version with respect to the mark 0. Let (V,N) be the stationary r.p.@ m.p.p. corresponding to (V_0^o, N_0^o) and (V_1^o, N_1^o) be the one corresponding to (V,N) Palm version with respect to the mark 1. Recall that $\tilde{V}$ is determined by $\tilde{N}_1^o$, $\{S_i^o\}$ and the initial condition. Hence

$$(5.1) \qquad \tilde{V}(t) \le \tilde{V}_1^o(t), \qquad t \in R.$$

Denote

$$\zeta = \inf\{t > 0: V_1^o(t) = 0\}.$$

Then

$$\tilde{V}(\zeta + t) = V_1^o(\zeta + t), \qquad t \in R_+,$$

and hence by the result of Example 3.11 and Lemma 1.2 the r.p. $\tilde{V}$ is stable.

Now consider the sequence $\{\tilde{V}_i^s\}$. Since (V_1^o, N_1^o) is a Palm version with respect to the mark 1 then $\{V_1^o(i+s), i \in Z\}$ is a stationary sequence. Clearly by (5.1)

$$(5.2) \qquad \tilde{V}(i+s) \le V_1^o(i+s), \qquad i \in Z.$$

If we denote

$$\zeta' = \inf\{i > 0: V_1^o(i+s) = 0\}$$

then recalling (5.2) we have

$$\tilde{V}(\zeta'+i+s) = V_1^o(\zeta'+i+s), \qquad i \in N_0.$$

Thus by Lemma 1.2 the sequence $\{\tilde{V}(i+s), i \in Z\}$ is strongly stable. Similarly we can show the stability of $\tilde{W}$. Namely define

$$W = \{W_i = V_1^o(Z_{1,i}^o - o), \quad i \in Z\}.$$

From (5.1)

$$\tilde{W}_i \le W_i , \qquad i \in Z .$$

Set

$$\zeta'' = \inf\{i > 0: W_i = 0\} .$$

Hence we obtain

$$\tilde{W}_{\zeta''+i} = W_{\zeta''+i} , \qquad i \in N_0 .$$

Unfortunatelly W is not a stationary sequence. Nevertheless we show that W is stable which in view of Lemma 1.2 yields the stability of $\tilde{W}$. Notice that

$$W = \varphi(V_1^o, N_1^o) ,$$

where $\varphi: \mathcal{D}(R, R_+) \times N_K \to R^\infty$ is a measurable mapping. An important property of φ is that

$$\theta^i \varphi = \varphi \sigma_0^i , \qquad i \in Z ,$$

where $\theta: R^Z \to R^Z$ is defined by

$$\theta(\{s_i , i \in Z\}) = \{s_{i+1}, i \in Z\} .$$

From Corollary 3.2 (ii)

$$\lim_{j\to\infty} \frac{1}{j} \sum_{i=0}^{j-1} \Pr(\sigma_0^i(V_1^o, N_1^o) \in F) = P_0^o(F) . \tag{5.3}$$

Hence for any $B \in \mathcal{B}R^\infty$

$$\lim_{j\to\infty} \frac{1}{j} \sum_{i=0}^{j-1} \Pr(\theta^i W \in B) = \lim_{j\to\infty} \frac{1}{j} \sum_{i=0}^{j-1} \Pr(\varphi(\sigma_0^i (V_1^o, N_1^o)) \in B) =$$

$$= \lim_{j\to\infty} \frac{1}{j} \sum_{i=0}^{j-1} \Pr(\sigma_0^i(V_1^o, N_1^o) \in \varphi^{-1} B) = P_0^o(\varphi^{-1} B) ,$$

which shows that $\tilde{W}$ is stable. The stability of $\tilde{L}$ follows from a result given in Example 3.14. This completes the proof of Proposition 5.1.

Denote by

F_V - the stationary d.f. of $\tilde{V}$,

F_{V^s} - the stationary d.f. of $\{\tilde{V}(i+s),\ i \in Z\}$,

F_W - the stationary d.f. of $\tilde{W}$,

F_L - the stationary d.f. of $\tilde{L}$.

The next proposition relates F_V, F_{V^s}, F_W, F_L. Denote $B(x) = \Pr(S_0^o \le x)$.

Proposition 5.2:

If $\rho < 1$ then

(the relation of the second type)

$$(5.4) \qquad F_V(t) = 1-\rho+\rho\, F_W * \tilde{B}(t) \ , \qquad t \in R_+$$

and (the Little formula)

$$(5.5) \qquad \int_{R_+} t\, F_L(dt) = \lambda\left(\int_{R_+} t\, F_W(dt) + \int_{R_+} t\, B(dt)\right) .$$

We also have

$$(5.6) \qquad F_V(t) = \int_0^1 F_{V^s}(t)ds \ .$$

Proof: Because of (5.3) it suffices to consider (V_0^o, N_0^o) and then applying the result proved in Example 3.18 we obtain (5.4). Similarly, by a result obtained in Example 3.10 the Little formula (5.5) follows. The formula (5.6) follows immediately from (3.8) namely

$$\Pr(V(o) \le t) = \int_0^1 \Pr(V_1^o(s) \le t)ds = \int_0^1 F_{V^s}(t)ds \ .$$

§ 3. Fagging queueing systems.

We consider now a single server, infinite capacity queue with the FIFO discipline. As in the foregoing examples it differs from a standard G/G/1; FIFO queue by another form of the generic process. We demonstrate that this system can be represented in a standard form of G/G/1; FIFO queue.

Suppose that within any busy cycle inter-arrival times and service times need not form a stationary sequence however the inter-arrival times and service times in different busy cycle follow the same probabilistic law according to a stationary scheme. To be more precise the generic sequence is given in the form of an array

$$\{(T^o_{ij}, S^o_{ij}),\ i \in Z,\ j \in N_0\} .$$

Rows form a metrically-transitive sequence of r.e.'s. For any $i \in Z$ let $\{V^o_i(t),\ t \in R_+\}$ be the v.w.t.p. in the queue with the generic sequence

$$\{(T^o_{ij}, S^o_{ij}),\ j \in N_0\}$$

and with the initial condition $V^o_i(0) = S^o_i$. The length of the first busy cycle of V^o_i we denote by τ^o_i . Denote

$$Y^o_{0,j} = \begin{cases} \sum_{i=0}^{j-1} \tau^o_i , & j=1,2,\ldots \\ 0 , & j=0 \\ -\sum_{i=-1}^{j} \tau^o_i , & j=-1,-2,\ldots . \end{cases}$$

Define the r.p. V^o by

$$V^o(t) = V^o_j(t - Y^o_{0,j}) , \qquad Y^o_{0,j} \le t < Y^o_{0,j+1} , \qquad t \in Z .$$

An m.p.p., with the space of marks $K = N_0$, associated with V^o is as follows. The i-th consecutive instant of the jump up of V^o in the j-th busy cycle is a point of N^o with the mark i $(i=0,1,\ldots)$. Its position is denoted by $Y^o_{i,j}$. The r.p.@ m.p.p. (V^o_0, N^o_0) is a Palm version with the mark 0 . For the existence of a stationary r.p.@ m.p.p. (V,N) corresponding to (V^o_0, N^o_0) we must assume

$$(5.7) \qquad E\,\tau^o_0 < \infty .$$

Denote by R^o_j the number of served customers in the j-th busy cycle, $j \in Z$. Set

$$\lambda_i = \frac{\Pr(R_0^o \geq i)}{E\,\tau_0^o} . \tag{5.8}$$

Proposition 5.3:

$$\lambda_i = EN([0,1] \times \{i\}) . \tag{5.9}$$

(λ_i is the intensity of points of N with the mark i).

Proof: Since rows are assumed independent, the r.p.@ m.p.p. (V_0^o, N_0^o) is $\underline{\sigma}_0$-ergodic. Hence by Lemma 3.2 (clearly $\lambda_0^{-1} = E\tau_0^o$) the r.p.@ m.p.p. (V,N) is ergodic. Thus

$$\lim_{t\to\infty} \frac{1}{t} N([0,t] \times \{i\}) = \lambda_i , \quad \text{a.e..} \tag{5.10}$$

Denote by P the distribution of N and by P_0^o the distribution of N_0^o. They are probability measures on $\mathcal{N}_{N_0}$. Set

$$B_i = \{n \in \mathcal{N}_{N_0} : \lim_{t\to\infty} \frac{1}{t} n([0,t] \times \{i\}) = \lambda_i\} .$$

Clearly B_i is an invariant set and from (4.20) we have $P(B_i) = 1$. Thus by Lemma 3.1 we get $P_0^o(B_i) = 1$. This yields

$$\lim_{t\to\infty} \frac{1}{t} N_0^o([0,t] \times \{i\}) = \lambda_i , \quad \text{a.e..} \tag{5.11}$$

On the other hand

$$\frac{N_0^o([0,t]\times\{i\})}{t} \leq \frac{\sum_{j=0}^{N_0^o([0,t]\times\{0\})} 1_{[i,\infty)}(R_j^o)}{N_0^o([0,t]\times\{0\})} \, \frac{N_0^o([0,t]\times\{0\})}{t}$$

and

$$\frac{\sum_{j=0}^{N_0^o([0,t)\times\{0\})-1} 1_{[i,\infty)}(R_j^o)}{N_0^o([0,t]\times\{0\})-1} \, \frac{N_0^o([0,t]\times\{0\})-1}{t} \leq \frac{N_0^o([0,t]\times\{i\})}{t} .$$

Thus by (5.11) we get (5.9).

If $\lambda_i > 0$ then there exists a Palm version (V_i^o, N_i^o) with respect to the mark i corresponding to (V,N). From Corollary 3.2 (ii)

$$F_{W_i}(x) = \Pr(V_i^o(0-o) \le x) = \lim_{l\to\infty} \frac{1}{l} \sum_{j=0}^{l-1} \Pr(V_0^o(Y_{i,j}^o - o) \le x) , \quad x \in R .$$

We also have by Corollary 3.2 (i)

$$F_V(x) = \Pr(V(0) \le x) = \lim_{t\to\infty} \frac{1}{t} \int_0^t \Pr(V_0^o(s) \le x) ds , \qquad x \in R .$$

Clearly F_{W_i} is the stationary d.f. of the i-th customer (i=0, ,...) in the busy cycle and F_V is the stationary d.f. of the virtual waiting time.

Theorem 5.1:

Assume that the array $\{S_{ij}^o\}$ consists of independent r.v.'s and $\{S_{ij}^o\}$ is independent of $\{T_{ij}^o\}$. Then

$$F_V(x) = 1 - \sum_{i=0}^{\infty} \rho_i + \sum_{i=0}^{\infty} \rho_i F_{W_i} * \tilde{B}_i(x) , \qquad x \in R_+ \tag{5.12}$$

where

$$\tilde{B}_i(x) = \frac{1}{ES_i^o} \int_0^x \Pr(S_i^o > t) dt ,$$

and

$$\rho_i = \lambda_i E S_i^o .$$

Proof: The proof follows from a result given in Example 3.18 applied to (V^o, N^o).

Corollary 5.1:

If moreover $\{T_{ij}^o\}$ consists of independent r.v.'s then

$$\Pr(V(0) = 0) = 1 - \frac{E \sum_{i=0}^{R_0^o} S_i^o}{E \sum_{i=0}^{R_0^o} T_i^o} .$$

Proof: From (4.12) we have $\Pr(V(0)=0) = F(0) = 1 - \sum_{i=0}^{\infty} \rho_i$. Now

$$\sum_{i=0}^{\infty} \rho_i = \frac{\sum_{i=0}^{\infty} ES_0^o \Pr(R_0^o \geq i)}{E\,\tau_0^o} .$$

Clearly $E\tau_0^o = E \sum_{i=0}^{R_0^o} T_i^o$. Since $1_{\{R_0^o < j\}}$ and S_j^o are independent, we have from the Kolmogorov-Prokhorov theorem

$$(5.13) \qquad E \sum_{i=0}^{R_0^o} S_i^o = \sum_{i=0}^{\infty} ES_i^o \Pr(R_0^o \geq i) .$$

§ 4. $\vec{G}/\vec{G}/1$ queue with work-conserving normal discipline.

In this section we consider a single server, infinite capacity queue with a fixed normal, work-conserving discipline. We dealt with such a queue in § 2.4 (i). The generic sequence is $\{T_i^o, S_i^o, K_i^o\}$ and is assumed metrically-transitive. We aim to study the following processes in the queue:

$\tilde{W} = \{\tilde{W}_i\}$ - the a.w.t.p.,

$\tilde{V} = \{\tilde{V}(t)\}$ - the v.w.t.p.,

$\tilde{L} = \{\tilde{L}(t)\}$ - the queue size process.

The associated m.p.p. is $N^o = \sum_i \delta_{(Y_i^o, K_i^o)}$ where

$$Y_j^o = \begin{cases} \sum_{i=0}^{j-1} T_i^o , & j=1,2,\ldots \\ 0 , & j=0 , \\ -\sum_{i=-1}^{j} T_i^o , & j=-1,-2,\ldots . \end{cases}$$

Proposition 5.4:

If $\rho = ES_0^o/ET_0^o < 1$ then $\tilde{W}$, $\tilde{V}$ and $\tilde{L}$ are stable.

Proof: It was proved in Theorem 2.5 that $\tilde{W}$ is stable. To study the v.w.t.p. $\tilde{V}$, define an a.w.t.p. $\tilde{M}$ in the associated G/G/1; FIFO queue by

$$\tilde{M}_i = \begin{cases} 0 , & i=-1,-2,\ldots \\ \tilde{W}_0 , & i=0 \\ \max(0,\tilde{M}_{i-1} + S^o_{i-1} - T^o_{i-1}) , & i=1,2,\ldots . \end{cases}$$

Clearly $\tilde{M}$ is stable. Having $\tilde{M}$, we may define the v.w.t.p. V^o and $\tilde{V}$ as it was done in Example 3.3. The r.p.@ m.p.p. $(\tilde{V},N^o)$ is $\underline{\sigma}$-stable and (V^o,N^o) is $\underline{\sigma}$-ergodic. Thus by Theorem 3.3, we have that $\tilde{V}$ is stable. To study the queue size process $\tilde{L}$ define the mapping

$$L^{\vee}: N_{R_+\times K} \to \mathcal{D}(R,R_+)$$

by

$$L^{\vee}(n,t) = \#\{i: y_i \le t < y_i + v_i\} , \qquad t \in R ,$$

where

$$n = \sum_i \delta_{(y_i,[v_i,k_i])} .$$

Notice the important property that

$$\tau_1^t L^{\vee} = L^{\vee}\tau_2^t , \qquad t \in R ,$$

where $\underline{\tau}_1$ is the group of shift transformations of $\mathcal{D}(R,R_+)$ into itself and $\underline{\tau}_2$ is the group of shift transformations of $N_{R_+\times K}$ into itself. Thus

$$\tilde{L}(t) = L^{\vee}(\bar{N},t) , \qquad t \in R ,$$

where

$$\bar{N} = \sum_i \delta_{(Y^o_i,[W^o_i+S^o_i,K^o_i])} .$$

In view of Theorem 2.5, the m.p.p. $\bar{N}$ is stable with an ergodic stationary distribution. This, by Theorem 3.3 (i), yields that $\tilde{L}$ is stable.

Denote by

F_W - the stationary d.f. of $\tilde{W}$,

F_V - the stationary d.f. of $\tilde{V}$,

F_L - the stationary d.f. of $\tilde{L}$,

Let M^o be a stationary a.w.t.p. in the associated G/G/1; FIFO queue with the generic sequence $\{T^o_i,S^o_i,K^o_i\}$. Denote

$F_M(x) = \Pr(M_0^o \le x)$, $\qquad F_{M_i^o}(x) = \Pr(M_0^o \le x \mid K_0^o = i)$,

$B(x) = \Pr(S_0^o \le x)$, $x \in R_+$, $i \in K$.

Proposition 5.5:

If $\rho = ES_0^o/ET_0^o < 1$ then

$$(5.14) \qquad \int_0^\infty tF_L(dt) = \lambda \left(\int_0^\infty tF_W(dt) + \int_0^\infty tB(dt) \right) .$$

If also

$$\Pr(S_0^o \le t \mid \{(T_i^o, S_i^o, K_i^o),\ i \ne 0\},\ T_0^o, K_0^o) = B_{K_0^o}(t) , \qquad \text{a.e.}$$

then

$$(5.15) \qquad F_V(t) = 1 - \rho + \sum_{i \in K} F_{M_i^o} * \tilde{B}_i(t) , \qquad t \in R_+ ,$$

$$\rho_i = \lambda_i \int_0^\infty t\, B_i(dt) , \qquad \rho = \sum_{i \in K} \rho_i .$$

Proof: Let W^o be a stationary a.w.t.p.. The result of Example 3.10 applied to

$$\bar{N}^o = \sum_i \delta_{(Y_i^o, [W_i^o + S_i^o, K_i^o])}$$

yields (5.14). Recalling that the v.w.t.p. is identical for all work-conserving queues, (5.15) follows from the result of Example 3.18.

§ 5. Takács relation in G/G/1; FIFO queues.

Consider a G/G/1; FIFO queue with a generic sequence $\{T_i^o, S_i^o, K_i^o\}$. The notations and assumptions of Example 3.2 are in force. We aim to find the second type Takács relation without the assumption (3.64). Denote by

F_V - the stationary d.f. of $\tilde{V}$,

F_W - the stationary d.f. of $\tilde{W}$,

F_D - the stationary d.f. of $\{\tilde{W}_i + S_i^o\}$,

namely the stationary d.f. of the sojourn time,

λ - the intensity of arrivals.

Theorem 5.2:

If $\rho = ES_0^o/ET_0^o < 1$ then

$$F_V(x) = \begin{cases} 0 \ , & x < 0 \\ 1 - \rho + \int_0^x f(t)dt \ , & x \geq 0 \end{cases}$$

and

$$f(x) = \lambda(F_W(x) - F_D(x)) \ , \qquad x \geq 0 \ .$$

Proof: We apply Theorem 3.5 to (V^o, N^o) defined in Example 3.2. Denote by (V,N) a stationary r.p.@ m.p.p. corresponding to (V^o,N^o). From Theorem 3.5 it follows that F_V $(F_V(x) = \Pr(V(0) \leq x),\ x \in R)$ belongs to the domain M_π of U_π, where $\{\pi_t(x,E)\} = \{1_E([x-t]^+)\}$. We now prove that

$$F_V(x) = a + \int_0^x f(t)dt \ , \qquad x \geq 0 \ ,$$

where $0 \leq a \leq 1$ and $f \geq 0$. Let $C_K''(R_+,R_+)$ be the class of functions $\varphi \in C_K'(R_+,R_+)$ with supp $\varphi \subset (0,\infty)$, where $C_K'(R_+,R_+)$ was defined in Example 3.16. Since $F_V \in M_\pi$, we have for $\varphi \in C_K''(R_+,R_+)$ that

$$\int_0^\infty \varphi(y) U_\pi F_V(dy) =$$

$$= \lim_{t \downarrow 0} \frac{1}{t} \left(\int_0^\infty \int_0^\infty \varphi(y)\pi_t(x,dy)F_V(dx) - \int_0^\infty \varphi(y)F_V(dy) \right) =$$

$$= \lim_{t \downarrow 0} \frac{1}{t} \int_0^\infty \left(\frac{\varphi([y-t]^+) - \varphi(y)}{t} \right) F_V(dy) = \int_0^\infty \varphi'(y)F_V(dy) \ .$$

Thus for $x > 0$

$$U_\pi F(x) = \frac{d}{dx} F(x)$$

and hence

$$F_V(x) = a + \int_0^x f(t)dt \ .$$

By Theorem 3.5 we obtain

$$f(x) = \lambda(F_W(x) - F_D(x)), \qquad x \geq 0 .$$

Clearly $W_i^o \leq W_i^o + S_i^o$ which yields $F_W(x) - F_D(x) \geq 0$. Thus $f \geq 0$. Now we find that

$$\lambda \int_0^\infty (F_W(x) - F_D(x))dx = \lambda \, E \, S_0^o = \rho$$

and hence

$$F_V(x) = 1 - \rho + \int_0^x f(t)dt ; \qquad x \geq 0 .$$

This completes the proof.

Corollary 5.2:

$$\int_0^\infty xF_V(dx) = \frac{\lambda}{2}\left(\int_0^\infty x^2F_D(dx) - \int_0^\infty x^2F_W(dx)\right) .$$

Proof: Using the results of Theorem 5.2 we have

$$\int_\theta^\infty xF_V(dx) = \int_0^\infty xf(x)dx$$

$$= \lambda\left(\int_0^\infty x(1 - F_D(x))dx - \int_0^\infty x(1 - F_W(x))dx\right)$$

$$= \frac{\lambda}{2}\left(\int_0^\infty x^2F_D(dx) - \int_0^\infty x^2F_W(dx)\right) .$$

§ 6. Takács relations in GI/GI/1; FIFO queues.

Consider a GI/GI/1; FIFO queue with a generic sequence $\{T_i^o, S_i^o\}$. Denote

$$A(x) = Pr(T_0^o \leq x) , \quad B(x) = Pr(S_0^o \leq x) , \quad x \in R .$$

Let F_V, F_W, λ be the same as in § 5. A stability condition is assumed that

$$\rho = \frac{ES_0^o}{ET_0^o} < 1 .$$

Let $\{W_i^o\}$ be a stationary a.w.t.p.. Since the queue is of the GI/GI/1; FIFO type we have

$$\Pr(T_0^o \le x \mid W_0^o = y) = A(x) , \qquad x \in R .$$

Thus the first type relation is

$$(5.16) \qquad F_V(x) = \begin{cases} 0 , & x < 0 \\ \int_0^\infty F_W * B(x+t)\tilde{A}(dt) , & x \ge 0 . \end{cases}$$

This follows from Theorem 3.4. The second type relation is (see Example 3.18)

$$(5.17) \qquad F_V(x) = 1 - \rho + \rho F_W * \tilde{B}(x) , \qquad x \in R_+ .$$

We can look at (5.16) and (5.17) as a system of integral equations for F_V, F_W. I conjecture that there is a unique solution of the system (in the set of probability measures on R_+). In such a case we would have a way of finding the stationary d.f. of the virtual waiting time and the stationary d.f. of the actual waiting time.

Consider now the simplest case of an M/GI/1; FIFO queue. This is a GI/GI/1; FIFO queue with a Poissonian input. Then

$$\Pr(T_0^o \le x \mid W_0^o = y) = 1 - e^{-\lambda x} , \qquad x \in R_+ ,$$

and by Proposition 4.1 we have $F_V = F_W$. Thus from (5.17) we obtain that F_W fulfills the integral equation

$$F_W(x) = (1 - \rho)\delta_0(x) + \rho F_W * \tilde{B}(x) , \qquad x \in R .$$

This equation has the unique solution

$$F_W(x) = \sum_{i=0}^{\infty} (1 - \rho)\rho^i \, \tilde{B}^{*i}(x) , \qquad x \in R .$$

Now we find the inverse second type relation.

Proposition 5.6:

If the d.f. of the service time is absolutely continuous then

$$F_W(x) = m_A(h * F_V(x) - (1-\rho)h(x)) , \tag{5.18}$$

where

$$h(x) = \frac{d}{dx} \sum_{i=1}^{\infty} B^{*i}(x) = \frac{d}{dx} H(x).$$

Proof: Convoluting the second type relation with B^{*k} we obtain

$$B^{*k} * F_V(x) = (1-\rho)B^{*k}(x) + \rho F_W * \tilde{B} * B^{*k}(x) , \qquad k \in N_0 .$$

Hence

$$\left(\sum_{k=0}^{\infty} B^{*k}\right) * F_V(x) = (1-\rho)h(x) + \lambda \int_0^x F_W(t)dt$$

which yields (5.18).

Notes

In a case of a Poisson periodic input p.p., formulae (5.4), (5.6) were obtained by Harrison & Lemoine (1977). "Fagging" single server systems were considered by Szczotka (1974). There he established the result of our Corollary 5.1. Some classes of GI/GI/1 work-conserving queues were analysed by Stidham (1972). He also dealt with Little type formulae and the second type formula in terms of moments, however the regenerative processes method used there cannot be applied to the more general case of $\vec{G}/\vec{G}/1$ queues. Formulae (5.16), (5.17) were established by Takács (1955), (1963).

References

Arndt, F. and Franken, P. (1979) Construction of a class of stationary processes with applications in reliability. *Zast. Mat.* 16, 379-393.

Borovkov, A.A. (1972) *Stochastic Processes in Queueing Theory.* (in Russian) Nauka, Moskva.

Breiman, L. (1968) *Probability.* Addison-Wesley Publishing Company.

Brill, P.H. and Posner, M.J. (1977) Level crossing in point processes applied to queues: single-server case. *Operations Res.* 25, 662-674.

Chung, K,L. (1960) *Markov Chains with Stationary Transition Probabilities.* Springer-Verlag, Berlin.

Çinlar, E. and Pinsky, M. (1972) On dams with additive inputs and a general release rule. *J. Appl. Prob.* 9, 422-429.

Cohen, J.W. (1969) *The Single Server Queue.* North Holland, Amsterdam.

Crane, M.A. and Iglehart, D.L. (1974) Simulating stable stochastic systems I: General multiserver queues. *J. ACM.* 21, 103-113.

Daley, D.J. and Trengove, C.D. (1977) Bounds for mean waiting times in single-server queues: a survay. Preprint.

Doob, J.L. (1953) *Stochastic Processes.* John Wiley and Sons, New York.

Franken, P. (1976) Einige Anwendungen der Theorie zufälliger Punktprozesse in der Bedienungstheorie I.*Math. Nachr.* 70, 309-319.

Franken, P. (1978) A remark on the stationary availability. *Math. Operationsforsch. Statist., Ser. Optimization.* 9, 143-144.

Franken, P. and Streller, A.(1979) Generalized regenerative processes. *Theor. Probability Appl.* 24, 78-89.

Harrison, J.M. and Lemoine, A.J. (1977) Limit theorems for periodic queues. *J. Appl. Prob.* 14, 566-576,

Harrison, J.M. and Resnick, S.J. (1976) The stationary distribution and first exit probabilities of a storage process with general release rule. *Mathematics of Operations Research*, 1, 347-358.

Jankiewicz, M. (1979a) Extended piecewise Markov processes in continuous time. *Zast. Mat.* 16, 175-196.

Jankiewicz, M. (1979b) Extended piecewise Markov processes in discrete time. *Zast. Mat.* 16, 197-206.

Jankiewicz, M. and Rolski, T. (1977) Piecewise Markov processes on a general state space. *Zast. Mat.* 15, 421-435.

Kälahne, U. (1976) Existence, uniqueness and some invariance properties of stationary distributions for general single server queues. *Math. Operationsforsch. Statist.* 7, 557-576.

Kallenberg, O. (1976) *Random Measures.* Academic Press.

Kerstan, J., Matthes, K. and Mecke, J. (1974) *Unbegrentz teilbare Punktprozesse.* Akademie-Verlag, Berlin.

Klimov, G.P. (1979) *Procesy Obsługi Masowej.* (in Polish) WNT, Warszawa.

Kopocińska, I. (1977a) Piecewise Markov processes in discrete time and their certain extensions. *Zast. Mat.* 16, 23-38.

Kopocińska, I. (1977b) Uogólnione procesy przedziałami markowskie. (in Polish) *Matematyka Stosowana* 9, 117-122.

Kopocińska, I. and Kopociński, B. (1971) Queueing systems with feedback. *Bull. Acad. Polon. Sci., Ser. Sci. Math. Astronom. Phys.* 19, 397--401.

Kopocińska, I. and Kopociński, B. (1972) Queueing systems with feedback. *Zast. Mat.* 12, 374-384.

Kopociński, B. and Rolski, T. (1977) On the virtual waiting time in the GI/G/s queue. *Bull. Acad. Polon. Sci., Ser. Sci. Math. Astronom. Phys.* 25, 1279-1280.

König, D., Rolski, T., Schmidt, V. and Stoyan, D. (1978) Stochastic processes with imbedded marked point process (PMP) and their application in queueing. *Math. Operationsforsch. Statist., Ser. Optimization*, 9, 125-141.

König, D. and Schmidt, V. (1980a) Relationship between time- and customer-stationary characteristics of service systems. *Proceedings of the Symposium on Point Processes and Queueing Theory* - Keszthely (Hungary) North Holland Publ. Comp. Amsterdam 1980.

König, D. and Schmidt, V. (1980b) Imbedded and non-imbedded stationary characteristics of queueing systems with varying service rate and point processes. *J. Appl. Probability* 17.

König, D. and Schmidt, V. (1980c) Stochastic inequalities between customer-stationary and time stationary characteristics of queueing systems with point processes. *J. Appl. Probability* 17.

Krakowski, M. (1973) Conservation methods in queueing theory. *Rev. franc. automat. infor. rech. oper.* 7, 63-83.

Kuczura, A. (1973) Piecewise Markov processes. *SIAM J. Appl. Math.* 24, 169-181.

Loynes, R.M. (1962) The stability of a queue with non-independent inter--arrival and service times. *Proc. Camb. Phil. Soc.* 58, 497-520.

Mecke, J. (1967) Stationäre zufällige Masse auf lokalkompakten Abelschen Gruppen. *Z. Wahrscheinlichkeitstheorie verw. Geb.* 9, 36-58.

Miyazawa, M. (1976a) Conservation laws in queueing theory and their application to the finiteness of moments. Research Report on Information Sciences, B-28. Tokyo Institute of Technology.

Miyazawa, M. (1976b) On the role of exponential distributions in queueing models. Research Report on Information Sciences. Dept. of Inf. Sciences, Tokyo Institute of Technology.

Miyazawa, M. (1977) Time and customer processes in queues with stationary inputs. *J. Appl. Prob.* 14, 349-357.

Mori, M. (1975) Some bounds for queues. *J. Operations Research Soc. of Japan*, 18, 152-181.

Papangelou, F. (1974) On the Palm probabilities of processes of points and processes of lines. In *Stochastic Geometry* (Harding, E,F. and Kendall, D.G. eds.) 114-147, John Wiley and Sons.

Rolski, T. (1976) Order relations in the set of probability distribution functions and their applications in queueing theory. *Dissertationes Math.* 132.

Rolski, T. (1977a) A relation between imbedded Markov chains in piecewise Markov processes. *Bull.Acad.Polon.Sci., Ser.Math.Astronom.Phys.* 25, 185-193.

Rolski, T. (1977b) On some classes of distribution functions determined by an order relation. *Proc. Symp. to Honour Jerzy Neyman*. PWN Warszawa, 293-302.

Rolski, T. (1978) A rate-conservative principle for stationary piecewise Markov processes. *Adv. Appl. Prob.* 10, 392-410.

Rolski, T. (1979) A note on queues with a common traffic intensity. *Math. Operationsforsch. Statist., Ser. Optimization* 10, 413-419.

Rudin, W. (1964) *Principles of Mathematical Analysis*. 2 nd ed., McGraw--Hill Book Company, New York.

Ryll-Nardzewski, C. (1961) Remarks on processes of calls. *Proc. of the 4-th Berkeley Symp. on Math. Stat. and Prob.* vol. 2, 455-465.

Schmidt, V. (1978) On some relations between stationary time and customer state probabilities for queueing systems G/GI/s/r. *Math. Operationsforsch. Statist., Ser. Optimization* 9, 261-272.

Stidham, S. Jr. (1972) Regenerative processes in the theory of queues with applications to alternating-priority queue. *Adv. Appl. Prob.* 4, 542-577.

Stoyan, D. (1977a) *Qualitative Eigenschaften und Abschatzungen stochastischer Modelle*. Akademie-Verlag, Berlin.

Stoyan, D. (1977b) Further stochastic order relations among GI/G/1 queues with a common traffic intensity. *Math. Operationsforsch. Statist., Ser. Optimization*. 8, 541-548.

Szczotka, W. (1974) Immediate service in a Beneš-type G/G/1 queueing systems. *Zast. Mat.* 14, 357-363.

Takács, L. (1955) Investigation of waiting time problems by reduction to Markov processes. *Acta Math. Acad. Sci. Hungaricae*. 6, 101-130.

Takács, L. (1963) The limiting distribution of the virtual waiting time and the queue size for a single-server queue with recurrent input and general service times. Sankhya A, 25, 91-100.

Whitt, W. (1972) Embedded renewal processes in the GI/G/s queue. *J. Appl. Prob.* 9, 650-658.

Whitt, W. (1979) Some useful functions for functional limit theorems. *Mathematics of Operations Research*.

Wolf, R. (1970) Work-conserving priorities. *J. Appl. Prob.* 7, 327-337.

Zähle, M. (1980) Ergodic properties of random fields and random geometric figures in the n-dimensional space with embedded point processes. Friedrich-Shiller-Universität Jena, Forschungsergebnisse. N/80/13.

INDEX

The following abbreviations are used: c.t. - continuous time, d.t. - discrete time.

Lecture Notes in Statistics

Vol. 1: R. A. Fisher: An Appreciation. Edited by S. E. Fienberg and D. V. Hinkley. xi, 208 pages, 1980.

Vol. 2: Mathematical Statistics and Probability Theory. Proceedings 1978. Edited by W. Klonecki, A. Kozek, and J. Rosiński. xxiv, 373 pages, 1980.

Vol. 3: B. D. Spencer, Benefit-Cost Analysis of Data Used to Allocate Funds. viii, 296 pages, 1980.

Vol. 4: E. A. van Doorn: Stochastic Monotonicity and Queueing Applications of Birth-Death Processes. vi, 118 pages, 1981.

Vol. 5: T. Rolski, Stationary Random Processes Associated with Point Processes. vi, 139 pages, 1981.

Vol. 6: S. S. Gupta and D.-Y. Huang, Multiple Statistical Decision Theory: Recent Developments. viii, 104 pages, 1981.

Vol. 7: M. Akahira and K. Takeuchi, Asymptotic Efficiency of Statistical Estimators. viii, 242 pages, 1981.

Vol. 8: The First Pannonian Symposium on Mathematical Statistics. Edited by P. Révész, L. Schmetterer, and V. M. Zolotarev. vi, 308 pages, 1981.

Springer Series in Statistics

L. A. Goodman and W. H. Kruskal, Measures of Association for Cross Classifications. x, 146 pages, 1979.

J. O. Berger, Statistical Decision Theory: Foundations, Concepts, and Methods. xiv, 420 pages, 1980.

R. G. Miller, Jr., Simultaneous Statistical Inference, 2nd edition. 300 pages, 1981.

P. Brémaud, Point Processes and Queues: Martingale Dynamics. 352 pages, 1981.

Lecture Notes in Mathematics

Selected volumes of interest to statisticians and probabilists:

Vol. 532: Théorie Ergodique. Actes des Journées Ergodiques, Rennes 1973/1974. Edited by J.-P. Conze, M. S. Keane. 227 pages, 1976.

Vol. 539: Ecole d'Ete de Probabilités de Saint-Flour V–1975. A. Badrikian, J. F. C. Kingman, J. Kuelbs. Edited by P.-L. Hennequin. 314 pages, 1976.

Vol. 550: Proceedings of the Third Japan-USSR Symposium on Probability Theory. Edited by G. Maruyama, J. V. Prokhorov. 722 pages, 1976.

Vol. 566: Empirical Distributions and Processes. Selected Papers from a Meeting at Oberwolfach, March 28–April 3, 1976. Edited by P. Gänssler, P. Revesz. 146 pages, 1976.

Vol. 581: Séminaire de Probabilités XI. Université de Strasbourg. Edited by C. Dellacherie, P. A. Meyer, and M. Weil. 573 pages, 1977.

Vol. 595: W. Hazod, Stetige Faltungshalbgruppen von Warscheinlichkeitsmassen und erzeugende Distributionen. 157 pages, 1977.

Vol. 598: Ecole d'Ete de Probabilités de Saint-Flour VI–1976. J. Hoffmann-Jorgensen, T. M. Ligett, J. Neveu. Edited by P.-L. Hennequin. 447 pages, 1977.

Vol. 636: Journées de Statistique des Processus Stochastiques, Grenoble 1977. Edited by Didier Dacunha-Castelle, Bernard Van Cutsem. 202 pages, 1978.

Vol. 649: Séminaire de Probabilités XII. Strasboug 1976–1977. Edited by C. Dellacherie, P. A. Meyer, and M. Weil. 805 pages, 1978.

Vol. 656: Probability Theory on Vector Spaces, Proceedings 1977. Edited by A. Weron. 274 pages. 1978.

Vol. 672: R. L. Taylor, Stochastic Convergence of Weighted Sums of Random Elements in Linear Spaces. 216 pages, 1978.

Vol. 675: J. Galambos, S. Kotz, Characterizations of Probability Distributions. 169 pages, 1978.

Vol. 678: Ecole d'Ete de Probabilités de Saint-Flour VII–1977. D. Dacunha-Castelle, H. Heyer, and B. Roynette. Edited by P.-L. Hennequin. 379 pages, 1978.

Vol. 690: W. J. J. Rey, Robust Statistical Methods. 128 pages, 1978.

Vol. 706: Probability Measures on Groups, Proceedings 1978. Edited by H. Heyer. 348 pages, 1979.

Vol. 714: J. Jacod, Calcul Stochastique et Problemes de Martingales. 539 pages, 1979.

Vol. 721: Séminaire de Probabilités XIII. Proceedings, Strasbourg, 1977/78. Edited by C. Dellacherie, P. A. Meyer, and M. Weil. 647 pages, 1979.

Vol. 794: Measure Theory, Oberwolfach 1979, Proceedings, 1979. Edited by D. Kolzöw. 573 pages, 1980.

Vol. 796: C. Constantinescu, Duality in Measure Theory. 197 pages, 1980.

Vol. 821: Statistique non Paramétrique Asymptotique, Proceedings 1979. Edited by J.-P. Raoult. 175 pages, 1980.

Vol. 828: Probability Theory on Vector Spaces II, Proceedings 1979. Edited by A. Weron. 324 pages, 1980.